Information Theory
of The Cosmos

Publishing-in-support-of,

EDUCREATION PUBLISHING

RZ 94, Sector - 6, Dwarka, New Delhi - 110075
Shubham Vihar, Mangla, Bilaspur, Chhattisgarh - 495001

Website: *www.educreation.in*

ISBN: 978-1-5457-1880-3
Price: ₹ 1,614.00

Printed in India

Information Theory of The Cosmos

How The New Science of Information Changes Our Understanding And Perception of The Universe

Raghunath Tiruvaipati

EDUCREATION PUBLISHING
(Since 2011)
www.educreation.in

*To My Mother
And Father*

Content List

Note: Page no. will be filled after finalisation.

Prologue

Information theory is a relatively new paradigm even though scientific research about it began in 1948 itself. it is synthesized from information revolution which is the greatest technological advancement in the 21st century. Computers , hard drives and internet have revolutionized the world. In 20th century itself, Communications technology has advanced from wired to wireless. Computers and communication technology are only tiny aspects of the information theory that we are going to talk about in this book. It deals not only with the way that machines like computers, cell-phones work , but also with objects on the scale from quantum level such as electron, photons, atoms etc.. to astronomical level such as planets, suns, galaxies , black holes and the universe itself. Its rules determine the life course of the universe and the entire cosmos as well. Information theory took birth or rather uncovered in the 20th century which has changed the world dramatically.

Information theory spans across thermo dynamics, theory of relativity , quantum mechanics, genetics, cosmology, astronomy and quantum computing. So I will touch upon all of these topics in this book.

Thermo dynamics - which explains about the working of engines, exchange of heat energy and motion of atoms is an information theory.

Theory of relativity – which explains about the large objects moving at extremely high speeds (almost at speed of light) and the influence on them by gravity is also an information theory.

Quantum mechanics – which explains about the domain of atoms and sub-atomic particles is also an information theory.

Genetics – genes are literally information-carrying bio-devices. So genetics is also an information theory. We will discuss it later.

Astronomy – all the galaxies, stars, planets contain a huge amount of information in the form of electromagnetic waves. And through these waves, they influence with each other. So astronomy is also a information theory.

Information is much more than what we assume about it (that is, simply some content) and co-relates all the above mentioned theories together.

Information theory changes our perception of the world radically because information is not something that is non-objective. Everything in the universe is constructed from information. So it is a rather important aspect of energy and matter. It is physical and can be quantized and measured which is contrary to our usual thinking. It is influenced by several physical laws which prescribe how it can be manipulated, transferred, copied, erased or destroyed. And everything in the universe must accord to the laws of the information theory.

As a result of the information theory, communications and computers have been transformed in a huge way. The laws of the information theory influences not only the data in a computer (data - bits and bytes which are ultimately the electrical voltage),but also codes and communication channels. They govern the way how the sub-atomic particles (electrons, quarks,etc..) behave . Even life obeys these laws and so does the entire universe.

Everything in the universe including life-forms on our earth are bundled with information. The DNA molecule in the living beings contains a huge amount of information in the form of genes which works like a blue print for the construction of the organism. Our body and mind processes trillions of bits of information each second unconsciously. Every particle , be it a photon, or an electron or a quark or an atom, is teeming with information in the form of spin, charge, size, electron configuration and weight which can be processed and transformed in energy transactions (chemical reactions). All the stars and galaxies in the universe are emitting information in the form of electro-magnetic waves which is spreading all over the cosmos.

Information seems to be the driving force behind the physical structure of the universe. It appears to solve the big scientific questions such as the origin of life, the black holes, the relativity theory and the quantum mechanics. The information theory is leading to a way to answer some of the most profound issues about the universe in the scientific community.

*****

Cryptography

Information should be transferred from place to place or from person to person in a secure way. The most ancient way to secure information is to use cryptography in the form of codes and ciphers. They must be secure enough so that when somebody taps the message, they must not be able to make it out.

Even though cryptography has been in place since time immemorial, the codes and ciphers were not secure enough. Somebody who has a little bit of knowledge about ciphers can break the encrypted message. Yet, people had to rely on these weak codes and ciphers to communicate sensitive information. Regardless of how well the information is encrypted, there is a chance for it to decrypted as it is passing from place to place. In the digital world, information is communicated in bits. To a sniffer, every bit is very much tangible and susceptible.

As said earlier, information is very much real and solid just as other properties of a particle. We can not see the spin, charge and electronic configuration of an atom directly but they are real. Similarly information is equally real. It can be quantized, evaluated and tampered with. Information has to travel physically from sender to receiver just like any other physical object. It can travel through a telephone line or a cable or wireless mechanism. Of course there is quantum entanglement in which two particles are entangled with each other over incredibly vast distance and still communicate with each other instantly. But the mechanism behind it is yet to be found out by the physicists.

Because information is a real entity, it can be displaced or sniffed through just like any other thing. But information must be propagated from place to place and from person to person to be of any value.

Now a days, encryption technology has advanced so much. A cryptographer has to design the cipher such that the information passes from the sender to the receiver without anybody sniffing that information in-between. The information essence of the encrypted message should not be perceived by anybody. Similarly , a decrypter tries to make out information from a random mixture of letter and symbols. He can break the encrypted message if the cipher is weak.

That the information is tangible and quantized is the main principle of the information theory. The science of cryptography will give an idea how information is tangible and how it must be propagated from place to place. One of the most important properties of information is its redundancy. If you grasp redundancy ,you can know why information is real and tangible.

When we receive an information, for example something like "this rose is beautiful", what is happening is that your brain is taking a series of words in the visual cortex and processing them to get the meaning of the message. when you read something on a paper, what is happening is you are receiving a series of symbols and try to grasp the meaning of that message. Your brain receives a set of symbols that say "this rose is beautiful" in the visual cortex and processes those symbols until its meaning is understood. All this happens at the sub-conscious level. This sub-conscious process is vital to our ability to use language. Redundancy is makes a message easy to grasp.

When a sentence is distorted , we can still decipher it. That is because of redundancy. Every sentence in any language is quite redundant and has more information than you need to make out its meaning. For example, take the sentence "j-st s—th-s" . it is quite distorted because all the vowels in the sentence are removed. How ever we can still decipher it and extract its meaning.

Redundancy is very useful. It allows for a sentence to be understood easily even if the sentence is distorted somewhat. Syntax and lexicographic rules create redundancy in all languages. When your are amidst a crowd, It is redundancy that helps us understand words spoken by others. Our brains unconsciously processes the words according to lexicographic rules.

Let us discuss the information redundancy in terms of computer science , communication theory and we will also discuss the enigma machine and the turing machine , two famous machines in history used for breaking the cryptgrams.

In computer science, information redundancy adds extra information to tolerate communication faults. Mainly two kinds of codes are used for fault tolerance : error detecting codes &

error correcting codes. These codes are used mostly in communication channels and computing science. Code of length n is a set of n-tuples satisfying some well-defined set of rules. Binary codes use only 0 and 1.

From the "Mathematical theory of information" by C.E.Shannon, the definition of information is as follows:

Definition1: information is the minimum number of bits, necessary to encode the subject message.
Definition2: information is the length of the shortest possible code, used to transport the message.

Example1: a report about rainfall , indicating weather there is rainfall or no rainfall alternatives, has information of 1 bit

Rainfall	code
Yes	1
No	0

Example2: information about a measurement scale- a more detailed one answering two simple yes/no questions: is it tall or short? And is it smooth or rough? The information content of the below table is 2 bits.

Tall?	Smooth?	code
Yes	Yes	11
Yes	No	10
No	Yes	01
No	No	00

If the message is longer than is necessary, that means if it is longer than information of the message, the code contains redundancy. It is the 2-logarithm of the quotient between the code length and the information of the message.

Redundancy=$\log_2$ (code length / information)

Redundancy can never be negative. It can be zero. when the code is as short as it can be (Code length is identical to the

INFORMATION, which is seldom the case). It is 1.0, when the code length is 2 x

INFORMATION. The REDUNDANCY present in different messages is eventually a compromise between
efficiency concerns and security issues. Example:

Measurement scale	code
Tall & smooth	TS
Tall & rough	TR
Short & smooth	SS
Short & rough	SR

The INFORMATION did not change and still amounts to 2 bits. The code length however increased from 2 to 16 bits.The REDUNDANCY of this code is $\log_2(16/2) = \log_2(8) = 3$.

Error detection and correction is one of the main concerns while transmitting information and it is characterized by the number of bits that can be corrected.

Double-bit detecting code can detect two single-bit errors

Single-bit detecting code can detect one single-bit error.

There is something called Hamming distance which gives the measure of error detection/correction capabilities of a code.

HAMMING DISTANCE

Hamming distance is the number of bit positions in which two n-tuples differ.

$$x\ 0000$$
$$y\ 0101$$

$$\delta(x.y) = 2$$

3-dimensional space (3-bit words)

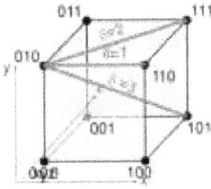

Error detection

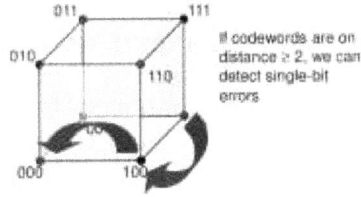

If codewords are on distance ≥ 2, we can detect single-bit errors

Error correction

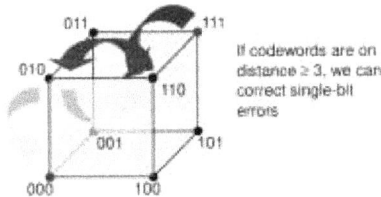

If codewords are on distance ≥ 3, we can correct single-bit errors

Code distance

Code distance is the minimum Hamming distance between any two distinct codewords.

$C_d = 2$ code detects all single-bit errors
code: 00, 11
invalid code words: 01 or 10

$C_d = 3$ code corrects all single-bit errors
code: 000, 111
invalid code words: 001, 010, 100,
101, 011, 110

Information and redundancy complement each other. When you deprive all the redundancy from a message, the left over is the information. Compression programs take this factor into account when downsizing the files.

The efficiency of a compression program is determined by the percentage of redundancy that it removes from the file. No matter how efficient the program is, there will be some data that is at the heart of the message. This chunk of information has to pass from sender to receiver.

Often cryptographers try to remove as much redundancy as possible. If he designs a good cipher, the message will be encrypted efficiently and it can be deciphered only by the intended recipient. Anybody who sniffs at the packet of information while it is passing from sender to receiver will not be able to make out the meaning of it. Often cipher exposes data because of redundancy.

The Enigma machine is a piece of hardware invented by a german and used by britain's code breakers as a way of deciphering german signals traffic during second world war.

A Turing machine is a hypothetical machine thought of by the mathematician Alan Turing in 1936. Despite its simplicity, the machine can simulate ANY computer algorithm, no matter how complicated it is!

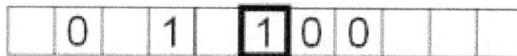

Above is a very simple representation of a Turing machine. It consists of an infinitely-long tape which acts like the memory in a typical computer, or any other form of data storage. The squares on the tape are usually blank at the start and can be written with symbols. In this case, the machine can only process the symbols 0 and 1 and " " (blank), and is thus said to be a 3-symbol Turing machine.

At any one time, the machine has a head which is positioned over one of the squares on the tape. With this head, the machine can perform three very basic operations:

Read the symbol on the square under the head.

Edit the symbol by writing a new symbol or erasing it.

Move the tape left of right by one square so that the machine can read and edit the symbol on a neighbouring square.

A simple demonstration

As a trivial example to demonstrate these operations, let's try printing the symbols "1 1 0" on an initially blank tape:

First, we write a 1 on the square under the head:

Next, we move the tape left by one square:

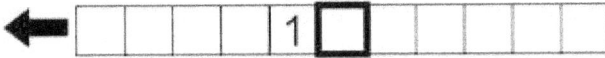

Now, write a 1 on the new square under the head:

We then move the tape left by one square again:

Finally, write a 0 and that's it!

A simple program

With the symbols "1 1 0" printed on the tape, let's attempt to convert the 1s to 0s and vice versa. This is called bit inversion, since 1s and 0s are bits in binary. This can be done by passing the following instructions to the Turing machine, utilising the machine's reading capabilities to decide its subsequent operations on its own. These instructions make up a simple program.

Symbol read	Write instruction	Move instruction
Blank	None	None
0	Write 1	Move tape to the right
1	Write 0	Move tape to the right

The machine will first read the symbol under the head, write a new symbol accordingly, then move the tape left or right as instructed, before repeating the read-write-move sequence again.

Let's see what this program does to our tape from the previous end point of the instructions:

The current symbol under the head is 0, so we write a 1 and move the tape right by one square.

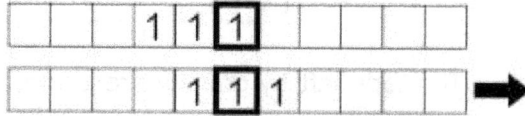

The symbol being read is now 1, so we write a 0 and move the tape right by one square:

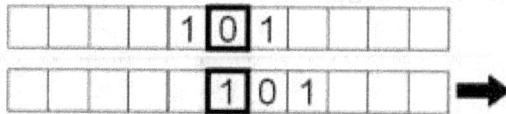

Similarly, the symbol read is a 1, so we repeat the same instructions.

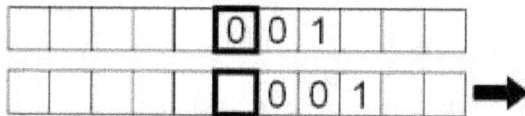

Finally, a 'blank' symbol is read, so the machine does nothing apart from read the blank symbol continuously since we have instructed it to repeat the read-write-move sequence without stopping.

In fact, the program is incomplete. How does the machine repeat the sequence endlessly, and how does the machine stop running the program? The program tells it to with the concept of a machine state.

The machine state

To complete the program, the state changes during the execution of the program on the machine must be considered. The following changes, marked in italics, must be aded to our table which can now be called a state table:

State	Symbol read	Write instruction	Move instruction	Next state
State 0	Blank	None	None	Stop state
	0	Write 1	Move the tape to the right	State 0
	1	Write 0	Move the tape to the right	State 0

We allocate the previous set of instructions to a machine state, so that the machine will perform those instructions when it is in the specified state.

After every instruction, we also specify a state for the machine to transition to. In the example, the machine is redirected back to its original state, State 0, to repeat the read-write-nove sequence, unless a blank symbol is read. When the machine reads a blank symbol, the machine is directed to a stop state and the program terminates.

Finite state machines

Even though it seems silly to do so, let's now add an additional state to our program that reverts the already inverted bits "1 1 0" back from "0 0 1" to "1 1 0". Below is the updated table, with changes listed in italics. The Turing machine now acts like a finite state machine with two states—these are called three-symbol, two-state Turing machines.

State	Symbol read	Write instruction	Move instruction	Next state
State 0	Blank	Write blank	Move the tape to the left	State 1
	0	Write 1	Move the tape to the right	State 1
	1	Write 0	Move the tape to the right	State 0
State 1	Blank	Write blank	Move the tape to the right	Stop state
	0	Write 1	Move the tape to the left	State 1
	1	Write 0	Move the tape to the left	State 1

For the write instruction, "None" has been changed to "Write blank" for uniformity's sake (so that only the machine's symbols are referred to), and it should be noted that they are equivalent.

Instead of a state table, the program can also be represented with a state diagram:

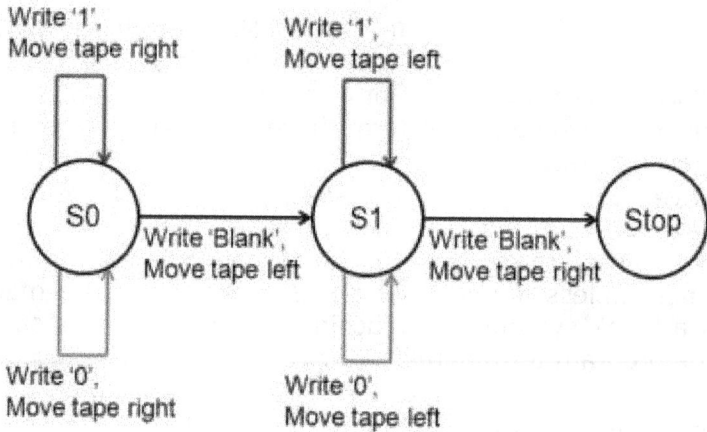

Write '1',
Move tape right

Write '1',
Move tape left

SO

Write 'Blank',
Move tape left

S1

Write 'Blank',
Move tape right

Stop

Write '0',
Move tape right

Write '0',
Move tape left

Legend:
⟶ Instruction when 'Blank' symbol is read
⟶ Instruction when '0' symbol is read
⟶ Instruction when '1' symbol is read

From where the program was previously, instead of doing nothing and stopping after the machine encounters a blank symbol, we instruct it to move the tape left before transitioning to State 1 where it reverses the bit inversion process.

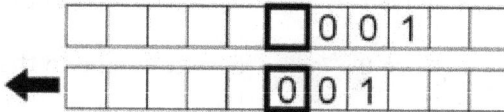

| | | | | | ☐ | 0 | 0 | 1 | | |

| | | | | | 0 | 0 | 1 | | |

Next, we invert the bits again, this time moving the tape left instead of right.

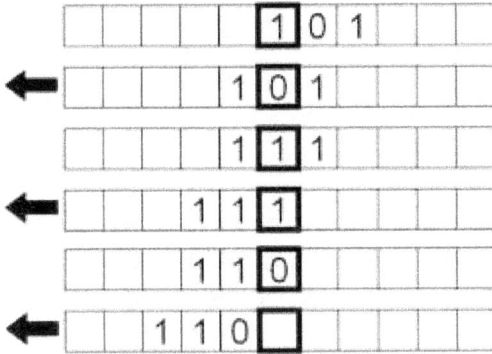

Finally, a blank symbol is read, so we move the tape right to get back to where we started, and stop the program.

With the introduction of more states to our program, we can instruct the Turing machine to perform more complex functions and hence run any algorithm that a modern day computer can.

*****

Thermodynamics

Boltzmann's equation is S=k logW. This equation has influenced both thermo dynamics as well as information theory. Initially it seems that there is no relationship between thermodynamics and information theory. Thermodynamics deals with heat, energy, and work where as information theory deals with information which is abstract and intangible most of the times. However information is rooted in the thermodynamics.

Boltzmann constant, (symbol k), a fundamental constant of physics occurring in nearly every statistical formulation of both classical and quantum physics. The constant is named after Ludwig Boltzmann, a 19th-century Austrian physicist, who substantially contributed to the foundation and development of statistical mechanics, a branch of theoretical physics. Having dimensions of energy per degree of temperature, the Boltzmann constant has a value of $1.38064852 \times 10^{-23}$ joule per kelvin (K), or $1.38064852 \times 10^{-16}$ erg per kelvin.

The physical significance of k is that it provides a measure of the amount of energy (i.e., heat) corresponding to the random thermal motions of the particles making up a substance. For a classical system at equilibrium at temperature T, the average energy per degree of freedom is kT/2. In the simplest example of a gas consisting of N non-interacting atoms, each atom has three translational degrees of freedom (it can move in the x-, y-, or z-directions), and so the total thermal energy of the gas is 3NkT/2.

When you burn something, the mass of the products is always equivalent to the mass of the reactants. this principle is known as conservation of mass. The law of conservation of mass or principle of mass conservation states that for any system closed to all transfers of matter and energy, the mass of the system must remain constant over time, as system's mass cannot change, so quantity cannot be added nor removed. Hence, the quantity of mass is conserved over time.

The law implies that mass can neither be created nor destroyed, although it may be rearranged in space, or the entities associated with it may be changed in form. For example, in chemical reactions, the mass of the chemical components before the reaction is equal to the mass of the components after the reaction. Thus, during any chemical reaction and low-energy thermodynamic processes in an isolated system, the total mass of the reactants, or starting materials, must be equal to the mass of the products.

The concept of mass conservation is widely used in many fields such as chemistry, mechanics, and fluid dynamics. Historically, mass conservation was discovered in chemical reactions independently by Mikhail Lomonosov and later rediscovered by Antoine Lavoisier in the late 18th century. The formulation of this law was of crucial importance in the progress from alchemy to the modern natural science of chemistry.

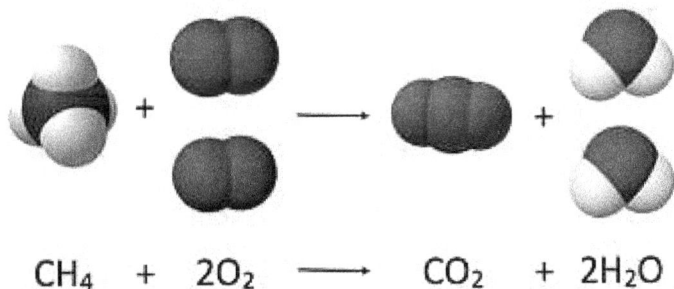

$$CH_4 + 2O_2 \longrightarrow CO_2 + 2H_2O$$

Combustion reaction of methane. Where 4 atoms of hydrogen, 4 atoms of oxygen and 1 of carbon are present before and after the reaction. The total mass after the reaction is the same as before the reaction.

The conservation of mass only holds approximately and is considered part of a series of assumptions coming from classical mechanics. The law has to be modified to comply with the laws of quantum mechanics and special relativity under the principle of mass-energy equivalence, which states that energy and mass form one conserved quantity. For very energetic systems the conservation of mass-only is shown not to hold, as is the case in nuclear reactions and particle-antiparticle annihilation in particle physics.

Mass is also not generally conserved in open systems. Such is the case when various forms of energy and matter are allowed into, or out of, the system. However, unless radioactivity or nuclear reactions are involved, the amount of energy escaping (or entering) such systems as heat, mechanical work, or electromagnetic radiation is usually too small to be measured as a decrease in system mass.

For systems where large gravitational fields are involved, general relativity has to be taken into account, where mass-energy conservation becomes a more complex concept, subject to different definitions, and neither mass nor energy is as strictly and simply conserved as is the case in special relativity.

Industrialization in Europe led to more powerful machines which required engines. James watt created a sophisticated steam engine.A steam engine is a heat engine that performs mechanical work using steam as its working fluid.

Steam engines are external combustion engines where the working fluid is separated from the combustion products. Non-combustion heat sources such as solar power, nuclear power or geothermal energy may be used. The ideal thermodynamic cycle used to analyze this process is called the Rankine cycle.

In the cycle, water is heated and changes into steam in a boiler operating at a high pressure. When expanded using pistons or turbines mechanical work is done. The reduced-pressure steam is then exhausted to the atmosphere, or condensed and pumped back into the boiler.

In general usage, the term steam engine can refer to either complete steam plants (including boilers etc.) such as railway steam locomotives and portable engines, or may refer to the piston or turbine machinery alone, as in the beam engine and stationary steam engine. However, a more detailed look at the steam locomotive referred to the engine as only that part where the heat in the steam was turned into motion of the piston, and hence enabled separate statements for boiler efficiency and engine efficiency. Specialized devices such as steam hammers and steam pile drivers are dependent on the steam pressure supplied from a separate boiler.

The use of boiling water to produce mechanical motion goes back over 2000 years, but early devices were not practical. The Spanish inventor Jerónimo de Ayanz y Beaumont obtained a patent for a rudimentary steamf-powered water pump in 1606. In 1698 Thomas Savery patented a steam pump that used steam in direct contact with the water being pumped. Savery's steam pump used condensing steam to create a vacuum and draw water into a chamber, and then applied pressurized steam to further pump the water.

Thomas Newcomen's atmospheric engine was the first commercial true steam engine using a piston, and was used in

1712 for pumping flood water from a mine. 104 were in use by 1733. Eventually over two thousand of them were installed.

In 1781 Scottish engineer James Watt patented a steam engine that produced continuous rotary motion. Watt's ten-horsepower engines enabled a wide range of manufacturing machinery to be powered. The engines could be sited anywhere that water and coal or wood fuel could be obtained. By 1883, engines that could provide 10,000 hp had become feasible. The stationary steam engine was a key component of the Industrial Revolution, allowing factories to locate where water power was unavailable. The atmospheric engines of Newcomen and Watt were large compared to the amount of power they produced, but high-pressure steam engines were light enough to be applied to vehicles such as traction engines and railway locomotives.

Reciprocating piston type steam engines remained the dominant source of power until the early 20th century, when advances in the design of electric motors and internal combustion engines gradually resulted in the replacement of reciprocating (piston) steam engines in commercial usage, and the ascendancy of steam turbines in power generation. Considering that the great majority of worldwide electric generation is produced by turbine type steam engines, the "steam age" is continuing with energy levels far beyond those of the turn of the 19th and 20th century.

Perpetual motion is motion of bodies that continues indefinitely. A perpetual motion machine is a hypothetical machine that can do work indefinitely without an energy source. This kind of machine is impossible, as it would violate the first or second law of thermodynamics.

These laws of thermodynamics apply even at very grand scales. For example, the motions and rotations of celestial bodies such as planets may appear perpetual, but are actually subject to many processes that slowly dissipate their kinetic energy, such as solar wind, interstellar medium resistance, gravitational radiation and thermal radiation, so they will not keep moving forever.

Thus, machines that extract energy from finite sources will not operate indefinitely, because they are driven by the energy stored in the source, which will eventually be exhausted. A common example is devices powered by ocean currents, whose energy is ultimately derived from the Sun, which itself will eventually burn out. Machines powered by more obscure sources have been proposed, but are subject to the same inescapable laws, and will eventually wind down.

In 2017 new states of matter, time crystals, were discovered in which on a microscopic scale the component atoms are in continual repetitive motion, thus satisfying the literal definition of "perpetual motion". However, these do not constitute perpetual motion machines in the traditional sense or violate thermodynamic laws because they are in their quantum ground state, so no energy can be extracted from them; they have "motion without energy".

In the philosophy of thermal and statistical physics, Maxwell's demon is a thought experiment created by the physicist James Clerk Maxwell in which he suggested how the second law of thermodynamics might hypothetically be violated. In the thought experiment, a demon controls a small door between two chambers of gas. As individual gas molecules approach the door, the demon quickly opens and shuts the door so that fast molecules pass into the other chamber, while slow molecules remain in the first chamber. Because faster molecules are hotter, the demon's behavior causes one chamber to warm up as the other cools, thus decreasing entropy and violating the second law of thermodynamics.

The thought experiment first appeared in a letter Maxwell wrote to Peter Guthrie Tait on 11 December 1867. It appeared again in a letter to John William Strutt in 1871, before it was presented to the public in Maxwell's 1872 book on thermodynamics titled Theory of Heat.

In his letters and books, Maxwell described the agent opening the door between the chambers as a "finite being." William Thomson (Lord Kelvin) was the first to use the word "demon" for Maxwell's concept, in the journal Nature in 1874, and implied that he intended the mediating, rather than malevolent, connotation of the word.

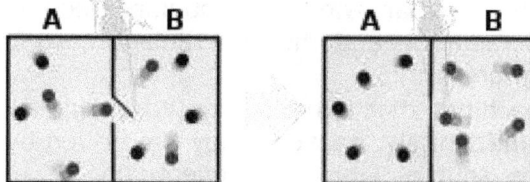

The second law of thermodynamics ensures (through statistical probability) that two bodies of different temperature, when brought into contact with each other and isolated from the rest of the Universe, will evolve to a thermodynamic equilibrium in which both bodies have approximately the same temperature. The second law is also expressed as the assertion that in an isolated system, entropy never decreases.

Maxwell conceived a thought experiment as a way of furthering the understanding of the second law. His description of the experiment is as follows:

... if we conceive of a being whose faculties are so sharpened that he can follow every molecule in its course, such a being, whose attributes are as essentially finite as our own, would be able to do what is impossible to us. For we have seen that molecules in a vessel full of air at uniform temperature are moving with velocities by no means uniform, though the mean velocity of any great number of them, arbitrarily selected, is almost exactly uniform. Now let us suppose that such a vessel is divided into two portions, A and B, by a division in which there is a small hole, and that a being, who can see the individual molecules, opens and closes this hole, so as to allow only the swifter molecules to pass from A to B, and only the slower molecules to pass from B to A. He will thus, without expenditure of work, raise the temperature of B and lower that of A, in contradiction to the second law of thermodynamics.

In other words, Maxwell imagines one container divided into two parts, A and B. Both parts are filled with the same gas at equal temperatures and placed next to each other. Observing the molecules on both sides, an imaginary demon guards a trapdoor between the two parts. When a faster-than-average molecule from A flies towards the trapdoor, the demon opens it, and the molecule will fly from A to B. Likewise, when a slower-than-average molecule from B flies towards the trapdoor, the demon will let it pass from B to A. The average speed of the molecules in B will have increased while in A they will have slowed down on average. Since average molecular speed corresponds to temperature, the temperature decreases in A and increases in B, contrary to the second law of thermodynamics. A heat engine operating between the thermal reservoirs A and B could extract useful work from this temperature difference.

The demon must allow molecules to pass in both directions in order to produce only a temperature difference; one-way passage only of faster-than-average molecules from A to B will cause higher temperature and pressure to develop on the B side.

*****

Information Theory

Information theory, a mathematical representation of the conditions and parameters affecting the transmission and processing of information. Most closely associated with the work of the American electrical engineer Claude Shannon in the mid-20th century, information theory is chiefly of interest to communication engineers, though some of the concepts have been adopted and used in such fields as psychology and linguistics. Information theory overlaps heavily with communication theory, but it is more oriented toward the fundamental limitations on the processing and communication of information and less oriented toward the detailed operation of particular devices.

Historical Background

Interest in the concept of information grew directly from the creation of the telegraph and telephone. In 1844 the American inventor Samuel F.B. Morse built a telegraph line between Washington, D.C., and Baltimore, Maryland. Morse encountered many electrical problems when he sent signals through buried transmission lines, but inexplicably he encountered fewer problems when the lines were suspended on poles. This attracted the attention of many distinguished physicists, most notably the Scotsman William Thomson (Baron Kelvin). In a similar manner, the invention of the telephone in 1875 by Alexander Graham Bell and its subsequent proliferation attracted further scientific notaries, such as Henri Poincaré, Oliver Heaviside, and Michael Pupin, to the problems associated with transmitting signals over wires. Much of their work was done using Fourier analysis, a technique described later in this article, but in all of these cases the analysis was dedicated to solving the practical engineering problems of communication systems.

The formal study of information theory did not begin until 1924, when Harry Nyquist, a researcher at Bell Laboratories, published a paper entitled "Certain Factors Affecting Telegraph

Speed." Nyquist realized that communication channels had maximum data transmission rates, and he derived a formula for calculating these rates in finite bandwidth noiseless channels. Another pioneer was Nyquist's colleague R.V.L. Hartley, whose paper "Transmission of Information" (1928) established the first mathematical foundations for information theory.

The real birth of modern information theory can be traced to the publication in 1948 of Claude Shannon's "A Mathematical Theory of Communication" in the Bell System Technical Journal. A key step in Shannon's work was his realization that, in order to have a theory, communication signals must be treated in isolation from the meaning of the messages that they transmit. This view is in sharp contrast with the common conception of information, in which meaning has an essential role. Shannon also realized that the amount of knowledge conveyed by a signal is not directly related to the size of the message. A famous illustration of this distinction is the correspondence between French novelist Victor Hugo and his publisher following the publication of Les Misérables in 1862. Hugo sent his publisher a card with just the symbol "?". In return he received a card with just the symbol "!". Within the context of Hugo's relations with his publisher and the public, these short messages were loaded with meaning; lacking such a context, these messages are meaningless. Similarly, a long, complete message in perfect French would convey little useful knowledge to someone who could understand only English.

Shannon thus wisely realized that a useful theory of information would first have to concentrate on the problems associated with sending and receiving messages, and it would have to leave questions involving any intrinsic meaning of a message—known as the semantic problem—for later investigators. Clearly, if the technical problem could not be solved—that is, if a message could not be transmitted correctly—then the semantic problem was not likely ever to be solved satisfactorily. Solving the technical problem was therefore the first step in developing a reliable communication system.

It is no accident that Shannon worked for Bell Laboratories. The practical stimuli for his work were the problems faced in creating a reliable telephone system. A key question that had to be answered in the early days of telecommunication was how best to maximize the physical plant—in particular, how to transmit the maximum number of telephone conversations over existing cables. Prior to Shannon's work, the factors for achieving maximum utilization were not clearly understood. Shannon's work defined communication channels and showed how to assign a capacity to them, not only in the theoretical sense where no interference, or noise, was present but also in practical cases where real channels were subjected to real noise. Shannon produced a formula that showed how the bandwidth of a channel (that is, its theoretical signal capacity) and its signal-to-noise ratio (a measure of interference) affected its capacity to carry signals. In doing so he was able to suggest strategies for maximizing the capacity of a given channel and showed the limits of what was possible with a given technology. This was of great utility to engineers, who could focus thereafter on individual cases and understand the specific trade-offs involved.

Shannon also made the startling discovery that, even in the presence of noise, it is always possible to transmit signals arbitrarily close to the theoretical channel capacity. This discovery inspired engineers to look for practical techniques to improve performance in signal transmissions that were far from optimal. Shannon's work clearly distinguished between gains that could be realized by adopting a different encoding scheme from gains that could be realized only by altering the communication system itself. Before Shannon, engineers lacked a systematic way of analyzing and solving such problems.

Shannon's pioneering work thus presented many key ideas that have guided engineers and scientists ever since. Though information theory does not always make clear exactly how to achieve specific results, people now know which questions are worth asking and can focus on areas that will yield the highest return. They also know which sorts of questions are difficult to answer and the areas in which there is not likely to be a large return for the amount of effort expended.

Since the 1940s and '50s the principles of classical information theory have been applied to many fields. The section Applications of information theory surveys achievements not only in such areas of telecommunications as data compression and error correction but also in the separate disciplines of physiology, linguistics, and physics. Indeed, even in Shannon's day many books and articles appeared that discussed the relationship between information theory and areas such as art and business. Unfortunately, many of these purported relationships were of dubious worth. Efforts to link information theory to every problem and every area were disturbing enough to Shannon himself that in a 1956 editorial titled "The Bandwagon" he issued the following warning:

I personally believe that many of the concepts of information theory will prove useful in these other fields—and, indeed, some results are already quite promising—but the establishing of such applications is not a trivial matter of translating words to a new domain, but rather the slow tedious process of hypothesis and experimental verification.

With Shannon's own words in mind, we can now review the central principles of classical information theory.

Classical Information Theory
Shannon's communication model

As the underpinning of his theory, Shannon developed a very simple, abstract model of communication, as shown in the figure. Because his model is abstract, it applies in many situations, which contributes to its broad scope and power.

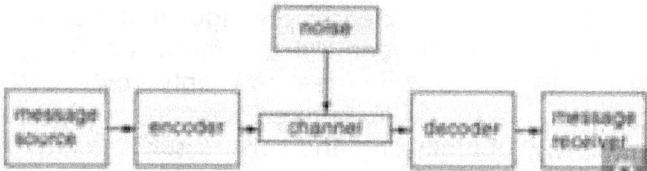

Shannon's communication model - Consider a simple telephone conversation: A person (message source) speaks into a telephone receiver (encoder), which converts the sound of the spoken word into an electrical signal. This electrical signal is then transmitted over telephone lines (channel) subject to interference (noise). When the signal reaches the telephone receiver (decoder) at the other end of the line it is converted back into vocal sounds. Finally, the recipient (message receiver) hears the original message.

The first component of the model, the message source, is simply the entity that originally creates the message. Often the message source is a human, but in Shannon's model it could also be an animal, a computer, or some other inanimate object. The encoder is the object that connects the message to the actual physical signals that are being sent. For example, there are several ways to apply this model to two people having a telephone conversation. On one level, the actual speech produced by one person can be considered the message, and the telephone mouthpiece and its associated electronics can be considered the encoder, which converts the speech into electrical signals that travel along the telephone network. Alternatively, one can consider the speaker's mind as the message source and the combination of the speaker's brain, vocal system, and telephone mouthpiece as the encoder. However, the inclusion of "mind" introduces complex semantic problems to any analysis and is generally avoided except for the application of information theory to physiology.

The channel is the medium that carries the message. The channel might be wires, the air or space in the case of radio and television transmissions, or fibre-optic cable. In the case of a signal produced simply by banging on the plumbing, the channel might be the pipe that receives the blow. The beauty of having an abstract model is that it permits the inclusion of a wide variety of channels. Some of the constraints imposed by channels on the propagation of signals through them will be discussed later.

Noise is anything that interferes with the transmission of a signal. In telephone conversations interference might be caused by static in the line, cross talk from another line, or background sounds. Signals transmitted optically through the air might suffer interference from clouds or excessive humidity. Clearly, sources of noise depend upon the particular communication system. A single system may have several sources of noise, but, if all of these separate sources are understood, it will sometimes be possible to treat them as a single source.

The decoder is the object that converts the signal, as received, into a form that the message receiver can comprehend. In the case of the telephone, the decoder could be the earpiece and its electronic circuits. Depending upon perspective, the decoder could also include the listener's entire hearing system.

The message receiver is the object that gets the message. It could be a person, an animal, or a computer or some other inanimate object.

Shannon's theory deals primarily with the encoder, channel, noise source, and decoder. As noted above, the focus of the theory is on signals and how they can be transmitted accurately and efficiently; questions of meaning are avoided as much as possible.

Four types of communication

There are two fundamentally different ways to transmit messages: via discrete signals and via continuous signals. Discrete signals can represent only a finite number of different, recognizable states. For example, the letters of the English alphabet are commonly thought of as discrete signals. Continuous signals, also known as analog signals, are commonly used to transmit quantities that can vary over an infinite set of values—sound is a typical example. However, such continuous quantities can be approximated by discrete signals—for instance, on a digital compact disc or through a digital telecommunication system—by increasing the number of distinct discrete values available until any inaccuracy in the description falls below the level of perception or interest.

Communication can also take place in the presence or absence of noise. These conditions are referred to as noisy or noiseless communication, respectively.

All told, there are four cases to consider: discrete, noiseless communication; discrete, noisy communication; continuous, noiseless communication; and continuous, noisy communication. It is easier to analyze the discrete cases than the continuous cases; likewise, the noiseless cases are simpler than the noisy cases. Therefore, the discrete, noiseless case will be considered first in some detail, followed by an indication of how the other cases differ.

Discrete, noiseless communication and the concept of entropy
From message alphabet to signal alphabet

As mentioned above, the English alphabet is a discrete communication system. It consists of a finite set of characters, such as uppercase and lowercase letters, digits, and various punctuation marks. Messages are composed by stringing these individual characters together appropriately. (Henceforth, signal components in any discrete communication system will be referred to as characters.)

For noiseless communications, the decoder at the receiving end receives exactly the characters sent by the encoder. However, these transmitted characters are typically not in the original message's alphabet. For example, in Morse Code appropriately spaced short and long electrical pulses, light flashes, or sounds are used to transmit the message. Similarly today, many forms of digital communication use a signal alphabet consisting of just two characters, sometimes called bits. These characters are generally denoted by 0 and 1, but in practice they might be different electrical or optical levels.

A key question in discrete, noiseless communication is deciding how to most efficiently convert messages into the signal alphabet. The concepts involved will be illustrated by the following simplified example.

The message alphabet will be called M and will consist of the four characters A, B, C, and D. The signal alphabet will be called S and will consist of the characters 0 and 1. Furthermore, it will be assumed that the signal channel can transmit 10 characters from S each second. This rate is called the channel capacity. Subject to these constraints, the goal is to maximize the transmission rate of characters from M.

The first question is how to convert characters between M and S. One straightforward way is shown in the table Encoding 1 of M using S. Using this conversion, the message ABC would be transmitted using the sequence 000110. The conversion from M to S is referred to as encoding. (This type of encoding is not meant to disguise the message but simply to adapt it to the nature of the communication system. Private or secret encoding schemes are usually referred to as encryption; see cryptology.) Because each character from M is represented by two characters from S and because the channel capacity is 10 characters from S each second, this communication scheme can transmit five characters from M each second. However, the scheme shown in the table ignores the fact that characters are used with widely varying frequencies in most alphabets.

Encoding 1 of M using S

M	→	S
A		00
B		01
C		10
D		11

In typical English text the letter e occurs roughly 200 times as frequently as the letter z. Hence, one way to improve the efficiency of the signal transmission is to use shorter codes for the more frequent characters—an idea employed in the design of Morse Code. For example, let it be assumed that generally one-half of the characters in the messages that we wish to send are the letter A, one-quarter are the letter B, one-eighth are the letter C, and one-eighth are the letter D. The table Encoding 2 of M using S summarizes this information and shows an alternative encoding for the alphabet M. Now the message ABC would be transmitted using the sequence 010110, which is also six characters long. To see that this second encoding is better, on average, than the first one requires a longer typical message. For instance, suppose that 120 characters from M are transmitted with the frequency distribution shown in this table.

Encoding 2 of M using S

frequency	M	→	S
50%	A		0
25%	B		10
12.5%	C		110
12.5%	D		111

The results are summarized in the table Comparison of two encodings from M to S. This table shows that the second encoding uses 30 fewer characters from S than the first encoding. Recall that the first encoding, limited by the channel capacity of 10 characters per second, would transmit five characters from M per second, irrespective of the message. Working under the same limitations, the second encoding would transmit all 120 characters from M in 21 seconds (210 characters from S at 10 characters per second)—which yields an average rate of about 5.7 characters per second. Note that this improvement is for a typical message (one that contains the expected frequency of A's and B's). For an atypical message—in this case, one with unusually many C's and D's—this encoding might actually take longer to transmit than the first encoding.

Comparison of two encodings from M to S

Character	no of cases	length of encoding1	length of encoding 2
A	60	120	60
B	30	60	60
C	15	30	45
D	15	30	45
Totals	120	240	210

A natural question to ask at this point is whether the above scheme is really the best possible encoding or whether something better can be devised. Shannon was able to answer this question using a quantity that he called "entropy"; his concept is discussed in a later section, but, before proceeding to that discussion, a brief review of some practical issues in decoding and encoding messages is in order.

Discrete, noisy communication and the problem of error

In the discussion above, it is assumed unrealistically that all messages are transmitted without error. In the real world, however, transmission errors are unavoidable—especially given the presence in any communication channel of noise, which is the sum total of random signals that interfere with the communication signal. In order to take the inevitable transmission errors of the real world into account, some adjustment in encoding schemes is necessary. The figure shows a simple model of transmission in the presence of noise, the binary symmetric channel. Binary indicates that this channel transmits only two distinct characters, generally interpreted as 0 and 1, while symmetric indicates that errors are equally probable regardless of which character is transmitted. The probability that a character is transmitted without error is labeled p; hence, the probability of error is 1 − p.

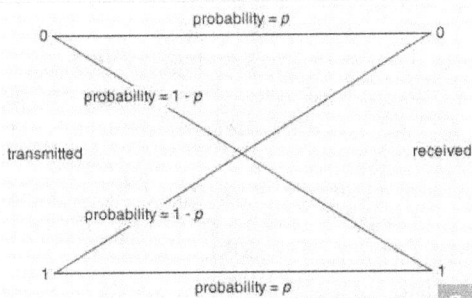

The binary symmetric channelThis type of channel transmits only two distinct characters, generally interpreted as 0 and 1, hence the designation binary. The probability of correctly receiving either character is the same, namely, p, which accounts for the designation symmetric.

Consider what happens as zeros and ones, hereafter referred to as bits, emerge from the receiving end of the channel. Ideally, there would be a means of determining which bits were received correctly. In that case, it is possible to imagine two printouts:

10110101010010011001010011101101000010100101—Signal
00000000000100000000100000000010000000011001—Errors

Signal is the message as received, while each 1 in Errors indicates a mistake in the corresponding Signal bit. (Errors itself is assumed to be error-free.)

Shannon showed that the best method for transmitting error corrections requires an average length of

$$E = p \log_2(1/p) + (1 - p) \log_2(1/(1 - p))$$

bits per error correction symbol. Thus, for every bit transmitted at least E bits have to be reserved for error corrections. A reasonable measure for the effectiveness of a binary symmetric channel at conveying information can be established by taking its raw throughput of bits and subtracting the number of bits necessary to transmit error corrections. The limit on the efficiency of a binary symmetric channel with noise can now be given as a percentage by the formula $100 \times (1 - E)$. Some examples follow.

Suppose that $p = 1/2$, meaning that each bit is received correctly only half the time. In this case $E = 1$, so the effectiveness of the channel is 0 percent. In other words, no information is being transmitted. In effect, the error rate is so high that there is no way to tell whether any symbol is correct—one could just as well flip a coin for each bit at the receiving end. On the other hand, if the probability of correctly receiving a character is .99, E is roughly .081, so the effectiveness of the channel is roughly 92 percent. That is, a 1 percent error rate results in the net loss of about 8 percent of the channel's transmission capacity.

One interesting aspect of Shannon's proof of a limit for minimum average error correction length is that it is nonconstructive; that is, Shannon proved that a shortest correction code must always exist, but his proof does not indicate how to construct such a code for each particular case. While Shannon's limit can always be approached to any desired degree, it is no trivial problem to find effective codes that are also easy and quick to decode.

Continuous communication and the problem of bandwidth

Continuous communication, unlike discrete communication, deals with signals that have potentially an infinite number of different values. Continuous communication is closely related to discrete communication (in the sense that any continuous signal can be approximated by a discrete signal), although the relationship is sometimes obscured by the more sophisticated mathematics involved.

The most important mathematical tool in the analysis of continuous signals is Fourier analysis, which can be used to model a signal as a sum of simpler sine waves. The figure indicates how the first few stages might appear. It shows a square wave, which has points of discontinuity ("jumps"), being modeled by a sum of sine waves. The curves to the right of the square wave show what are called the harmonics of the square wave. Above the line of harmonics are curves obtained by the addition of each successive harmonic; these curves can be seen to resemble the square wave more closely with each addition. If the entire infinite set of harmonics were added together, the square wave would be reconstructed almost exactly. Fourier analysis is useful because most communication circuits are linear, which essentially means that the whole is equal to the sum of the parts. Thus, a signal can be studied by separating, or decomposing, it into its simpler harmonics.

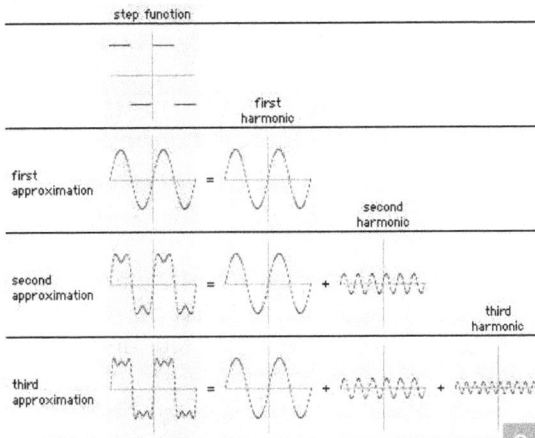

An example of Fourier analysis Using Fourier analysis, a step function is modeled, or decomposed, as the sum of various sine functions. This striking example demonstrates how even an obviously discontinuous and piecewise linear graph (a step function) can be reproduced to any desired level of accuracy by combining enough sine functions, each of which is continuous and nonlinear.

A signal is said to be band-limited or bandwidth-limited if it can be represented by a finite number of harmonics. Engineers limit the bandwidth of signals to enable multiple signals to share the same channel with minimal interference. A key result that pertains to bandwidth-limited signals is Nyquist's sampling theorem, which states that a signal of bandwidth B can be reconstructed by taking 2B samples every second. In 1924, Harry Nyquist derived the following formula for the maximum data rate that can be achieved in a noiseless channel:

Maximum Data Rate = $2\ B \log_2 V$ bits per second,

where B is the bandwidth of the channel and V is the number of discrete signal levels used in the channel. For example, to send only zeros and ones requires two signal levels. It is possible to envision any number of signal levels, but in practice the difference between signal levels must get smaller, for a fixed bandwidth, as the number of levels increases. And as the differences between signal levels decrease, the effect of noise in the channel becomes more pronounced.

Every channel has some sort of noise, which can be thought of as a random signal that contends with the message signal. If the noise is too great, it can obscure the message. Part of Shannon's seminal contribution to information theory was showing how noise affects the message capacity of a channel. In particular, Shannon derived the following formula:

Maximum Data Rate = $B \log_2(1 + S/N)$ bits per second,

where B is the bandwidth of the channel, and the quantity S/N is the signal-to-noise ratio, which is often given in decibels (dB). Observe that the larger the signal-to-noise ratio, the greater the data rate. Another point worth observing, though, is that the log2 function grows quite slowly. For example, suppose S/N is 1,000, then log2 1,001 = 9.97. If S/N is doubled to 2,000, then log2 2,001 = 10.97. Thus, doubling S/N produces only a 10 percent gain in the maximum data rate. Doubling S/N again would produce an even smaller percentage gain.

Applications Of Information Theory:
Data compression

Shannon's concept of entropy (a measure of the maximum possible efficiency of any encoding scheme) can be used to determine the maximum theoretical compression for a given message alphabet. In particular, if the entropy is less than the average length of an encoding, compression is possible.

The table Relative frequencies of characters in English text shows the relative frequencies of letters in representative English text. The table assumes that all letters have been capitalized and ignores all other characters except for spaces. Note that letter frequencies depend upon the particular text sample. An essay about zebras in the zoo, for instance, is likely to have a much greater frequency of z's than the table would suggest. Nevertheless, the frequency distribution for any very large sample of English text would appear quite similar to this table. Calculating the entropy for this distribution gives 4.08 bits per character. (Recall Shannon's formula for entropy.) Because normally 8 bits per character are used in the most common coding standard, Shannon's theory shows that there exists an encoding that is roughly twice as efficient as the normal one for this simplified message alphabet. These results, however, apply only to large samples and assume that the source of the character stream transmits characters in a random fashion based on the probabilities in the table. Real text does not perfectly fit this model; parts of it tend to be highly nonrandom and repetitive. Thus, the theoretical results do not immediately translate into practice.

Relative frequencies of characters in English text:
Character | relative frequency | character | relative frequency (relative frequency is mentioned in probabilities)

Character	relative frequency	character	relative frequency
(space)	.1859	F	.0208
E	.1031	M	.0198
T	.0796	W	.0175
A	.0642	Y	.0164
O	.0632	P	.0152
I	.0575	G	.0152
N	.0574	B	.0127
S	.0514	V	.0083
R	.0484	K	.0049
H	.0467	X	.0013
L	.0321	Q	.0008
D	.0317	J	.0008
U	.0228	Z	.0005
C	.0218		

In 1977–78 the Israelis Jacob Ziv and Abraham Lempel published two papers that showed how compression can be done dynamically. The basic idea is to store blocks of text in a dictionary and, when a block of text reappears, to record which block was repeated rather than recording the text itself. Although there are technical issues related to the size of the dictionary and the updating of its entries, this dynamic approach to compression has proved very useful, in part because the compression algorithm adapts to optimize the encoding based upon the particular text. Many computer programs use compression techniques based on these ideas. In practice, most text files compress by about 50 percent—that is, to approximately 4 bits per character. This is the number suggested by the entropy calculation.

Error-correcting and error-detecting codes
Shannon's work in the area of discrete, noisy communication pointed out the possibility of constructing error-correcting codes. Error-correcting codes add extra bits to help correct errors and thus operate in the opposite direction from

compression. Error-detecting codes, on the other hand, indicate that an error has occurred but do not automatically correct the error. Frequently the error is corrected by an automatic request to retransmit the message. Because error-correcting codes typically demand more extra bits than error-detecting codes, in some cases it is more efficient to use an error-detecting code simply to indicate what has to be retransmitted.

Deciding between error-correcting and error-detecting codes requires a good understanding of the nature of the errors that are likely to occur under the circumstances in which the message is being sent. Transmissions to and from space vehicles generally use error-correcting codes because of the difficulties in getting retransmission. Because of the long distances and low power available in transmitting from space vehicles, it is easy to see that the utmost skill and art must be employed to build communication systems that operate at the limits imposed by Shannon's results.

A common type of error-detecting code is the parity code, which adds one bit to a block of bits so that the ones in the block always add up to either an odd or even number. For example, an odd parity code might replace the two-bit code words 00, 01, 10, and 11 with the three-bit words 001, 010, 100, and 111. Any single transformation of a 0 to a 1 or a 1 to a 0 would change the parity of the block and make the error detectable. In practice, adding a parity bit to a two-bit code is not very efficient, but for longer codes adding a parity bit is reasonable. For instance, computer and fax modems often communicate by sending eight-bit blocks, with one of the bits reserved as a parity bit. Because parity codes are simple to implement, they are also often used to check the integrity of computer equipment.

As noted earlier, designing practical error-correcting codes is not easy, and Shannon's work does not provide direct guidance in this area. Nevertheless, knowing the physical characteristics of the channel, such as bandwidth and signal-to-noise ratio, gives valuable knowledge about maximum data transmission capabilities.

Cryptology
Cryptology is the science of secure communication. It concerns both cryptanalysis, the study of how encrypted information is revealed (or decrypted) when the secret "key" is unknown, and

cryptography, the study of how information is concealed and encrypted in the first place.

Shannon's analysis of communication codes led him to apply the mathematical tools of information theory to cryptography in "Communication Theory of Secrecy Systems" (1949). In particular, he began his analysis by noting that simple transposition ciphers—such as those obtained by permuting the letters in the alphabet—do not affect the entropy because they merely relabel the characters in his formula without changing their associated probabilities.

Cryptographic systems employ special information called a key to help encrypt and decrypt messages. Sometimes different keys are used for the encoding and decoding, while in other instances the same key is used for both processes. Shannon made the following general observation: "the amount of uncertainty we can introduce into the solution cannot be greater than the key uncertainty." This means, among other things, that random keys should be selected to make the encryption more secure. While Shannon's work did not lead to new practical encryption schemes, he did supply a framework for understanding the essential features of any such system.

Linguistics

While information theory has been most helpful in the design of more efficient telecommunication systems, it has also motivated linguistic studies of the relative frequencies of words, the length of words, and the speed of reading.

The best-known formula for studying relative word frequencies was proposed by the American linguist George Zipf in Selected Studies of the Principle of Relative Frequency in Language (1932). Zipf's Law states that the relative frequency of a word is inversely proportional to its rank. That is, the second most frequent word is used only half as often as the most frequent word, and the 100th most frequent word is used only one hundredth as often as the most frequent word.

Consistent with the encoding ideas discussed earlier, the most frequently used words tend to be the shortest. It is uncertain how much of this phenomenon is due to a "principle of least effort," but using the shortest sequences for the most common words certainly promotes greater communication efficiency.

Information theory provides a means for measuring redundancy or efficiency of symbolic representation within a given language. For example, if English letters occurred with equal regularity (ignoring the distinction between uppercase and lowercase letters), the expected entropy of an average sample of English text would be log2(26), which is approximately 4.7. The table Relative frequencies of characters in English text shows an entropy of 4.08, which is not really a good value for English because it overstates the probability of combinations such as qa. Scientists have studied sequences of eight characters in English and come up with a figure of about 2.35 for the average entropy of English. Because this is only half the 4.7 value, it is said that English has a relative entropy of 50 percent and a redundancy of 50 percent.

A redundancy of 50 percent means that roughly half the letters in a sentence could be omitted and the message still be reconstructable. The question of redundancy is of great interest to crossword puzzle creators. For example, if redundancy was 0 percent, so that every sequence of characters was a word, then there would be no difficulty in constructing a crossword puzzle because any character sequence the designer wanted to use would be acceptable. As redundancy increases, the difficulty of creating a crossword puzzle also increases. Shannon showed that a redundancy of 50 percent is the upper limit for constructing two-dimensional crossword puzzles and that 33 percent is the upper limit for constructing three-dimensional crossword puzzles.

Shannon also observed that when longer sequences, such as paragraphs, chapters, and whole books, are considered, the entropy decreases and English becomes even more predictable. He considered longer sequences and concluded that the entropy of English is approximately one bit per character. This indicates that in longer text nearly all of the message can be guessed from just a 20 to 25 percent random sample.

Various studies have attempted to come up with an information processing rate for human beings. Some studies have concentrated on the problem of determining a reading rate. Such studies have shown that the reading rate seems to be independent of language—that is, people process about the same number of bits whether they are reading English or Chinese. Note that although Chinese characters require more bits for their representation than English letters—there exist about 10,000 common Chinese characters, compared with 26 English letters—they also contain more information. Thus, on balance, reading rates are comparable.

Algorithmic information theory

In the 1960s the American mathematician Gregory Chaitin, the Russian mathematician Andrey Kolmogorov, and the American engineer Raymond Solomonoff began to formulate and publish an objective measure of the intrinsic complexity of a message. Chaitin, a research scientist at IBM, developed the largest body of work and polished the ideas into a formal theory known as algorithmic information theory (AIT). The algorithmic in AIT comes from defining the complexity of a message as the length of the shortest algorithm, or step-by-step procedure, for its reproduction.

Physiology

Almost as soon as Shannon's papers on the mathematical theory of communication were published in the 1940s, people began to consider the question of how messages are handled inside human beings. After all, the nervous system is, above all else, a channel for the transmission of information, and the brain is, among other things, an information processing and messaging centre. Because nerve signals generally consist of pulses of electrical energy, the nervous system appears to be an example of discrete communication over a noisy channel. Thus, both physiology and information theory are involved in studying the nervous system.

Many researchers (being human) expected that the human brain would show a tremendous information processing capability. Interestingly enough, when researchers sought to measure information processing capabilities during "intelligent" or "conscious" activities, such as reading or piano playing, they came up with a maximum capability of less than 50 bits per second. For example, a typical reading rate of 300 words per minute works out to about 5 words per second. Assuming an average of 5 characters per word and roughly 2 bits per character yields the aforementioned rate of 50 bits per second. Clearly, the exact number depends on various assumptions and could vary depending on the individual and the task being performed. It is known, however, that the senses gather some 11 million bits per second from the environment.

The table Information transmission rates of the senses shows how much information is processed by each of the five senses. This table immediately directs attention to the problem of determining what is happening to all this data. In other words, the human body sends 11 million bits per second to the brain for processing, yet the conscious mind seems to be able to process only 50 bits per second.

Information transmission rates of the senses
sensory system   bits per second

sensory system	bits per second
eyes	10,000,000
skin	1,000,000
ears	100,000
smell	100,000
taste	1,000

It appears that a tremendous amount of compression is taking place if 11 million bits are being reduced to less than 50. Note that the discrepancy between the amount of information being transmitted and the amount of information being processed is so large that any inaccuracy in the measurements is insignificant.

Two more problems suggest themselves when thinking about this immense amount of compression. First is the problem of determining how long it takes to do the compression, and second is the problem of determining where the processing power is found for doing this much compression.

The solution to the first problem is suggested by the approximately half-second delay between the instant that the senses receive a stimulus and the instant that the mind is conscious of a sensation. (To compensate for this delay, the body has a reflex system that can respond in less than one-tenth of second, before the mind is conscious of the stimulus.) This half-second delay seems to be the time required for processing and compressing sensory input.

The solution to the second problem is suggested by the approximately 100 billion cells of the brain, each with connections to thousands of other brain cells. Equipped with this many processors, the brain might be capable of executing as many as 100 billion operations per second, a truly impressive number.

It is often assumed that consciousness is the dominant feature of the brain. The brief observations above suggest a rather different picture. It now appears that the vast majority of processing is accomplished outside conscious notice and that most of the body's activities take place outside direct conscious control. This suggests that practice and habit are important because they train circuits in the brain to carry out some actions "automatically," without conscious interference. Even such a "simple" activity as walking is best done without interference from consciousness, which does not have enough information processing capability to keep up with the demands of this task.

The brain also seems to have separate mechanisms for short-term and long-term memory. Based on psychologist George Miller's paper "The Magical Number Seven, Plus or Minus Two: Some Limits on Our Capacity for Processing Information" (1956), it appears that short-term memory can only store between five and nine pieces of information to which it has been exposed only briefly. Note that this does not mean between five and nine bits, but rather five to nine chunks of information. Obviously, long-term memory has a greater capacity, but it is not clear exactly how the brain stores information or what limits may exist. Some scientists hope that information theory may yet afford further insights into how the brain functions.

Physics

The term entropy was originally introduced by the German physicist Rudolf Clausius in his work on thermodynamics in the 19th century. Clausius invented the word so that it would be as close as possible to the word energy. In certain formulations of statistical mechanics a formula for entropy is derived that looks confusingly similar to the formula for entropy derived by Shannon.

There are various intersections between information theory and thermodynamics. One of Shannon's key contributions was his analysis of how to handle noise in communication systems. Noise is an inescapable feature of the universe. Much of the noise that occurs in communication systems is a random noise, often called thermal noise, generated by heat in electrical circuits. While thermal noise can be reduced, it can never be completely eliminated. Another source of noise is the homogeneous cosmic background radiation, believed to be a remnant from the creation of the universe. Shannon's work permits minimal energy costs to be calculated for sending a bit of information through such noise.

Another problem addressed by information theory was dreamed up by the Scottish physicist James Clerk Maxwell in 1871. Maxwell created a "thought experiment" that apparently violates the second law of thermodynamics. This law basically states that all isolated systems, in the absence of an input of energy, relentlessly decay, or tend toward disorder. Maxwell began by postulating two gas-filled vessels at equal temperatures, connected by a valve. (Temperature can be defined as a measure of the average speed of gas molecules, keeping in mind that individual molecules can travel at widely varying speeds.) Maxwell then described a mythical creature, now known as Maxwell's demon, that is able rapidly to open and close the valve so as to allow only fast-moving molecules to pass in one direction and only slow-moving molecules to pass in the other direction. Alternatively, Maxwell envisioned his demon allowing molecules to pass through in only one direction. In either case, a "hot" and a "cold" vessel or a "full" and "empty" vessel, the apparent result is two vessels that, with no input of energy from an external source, constitute a more orderly isolated system—thus violating the second law of thermodynamics.

Information theory allows one exorcism of Maxwell's demon to be performed. In particular, it shows that the demon needs information in order to select molecules for the two different vessels but that the transmission of information requires energy. Once the energy requirement for collecting information is included in the calculations, it can be seen that there is no violation of the second law of thermodynamics.

*****

Genetic Code

Genetic code, the sequence of nucleotides in deoxyribonucleic acid (DNA) and ribonucleic acid (RNA) that determines the amino acid sequence of proteins. Though the linear sequence of nucleotides in DNA contains the information for protein sequences, proteins are not made directly from DNA. Instead, a messenger RNA (mRNA) molecule is synthesized from the DNA and directs the formation of the protein. RNA is composed of four nucleotides: adenine (A), guanine (G), cytosine (C), and uracil (U). Three adjacent nucleotides constitute a unit known as the codon, which codes for an amino acid. For example, the sequence AUG is a codon that specifies the amino acid methionine. There are 64 possible codons, three of which do not code for amino acids but indicate the end of a protein. The remaining 61 codons specify the 20 amino acids that make up proteins. The AUG codon, in addition to coding for methionine, is found at the beginning of every mRNA and indicates the start of a protein. Because most of the 20 amino acids are coded for by more than one codon, the code is called degenerate.

The genetic code, once thought to be identical in all forms of life, has been found to diverge slightly in certain organisms and in the mitochondria of some eukaryotes. Nevertheless, these differences are rare, and the genetic code is identical in almost all species, with the same codons specifying the same amino acids. The deciphering of the genetic code was accomplished by the American biochemists Marshall W. Nirenberg, Robert W. Holley, and Har Gobind Khorana in the early 1960s.

Nucleotide triplets (codons) specifying different amino acids are shown in the table.

The genetic code: Nucleotide triplets (codons) specifying different amino acids in protein chains*

*The columns may be read: the DNA triplet is transcribed into an RNA triplet, which then directs the production of an amino acid.

DNA triplet	RNA triplet	amino acid
AAA	UUU	phenylalanine
AAG	UUC	
AAT	UUA	leucine
AAC	UUG	
GAA	CUU	
GAG	CUC	
GAT	CUA	
GAC	CUG	
AGA	UCU	serine
AGG	UCC	
AGT	UCA	
AGC	UCG	
TCA	AGU	
TCG	AGC	
GGA	CCU	proline
GGG	CCC	
GGT	CCA	
GGC	CCG	
TAA	AUU	isoleucine (Ileu)
TAG	AUC	
TAT	AUA	
TAC	AUG	methionine
TGA	ACU	threonine
TGG	ACC	
TGT	ACA	
TGC	ACG	
CAA	GUU	valine
CAG	GUC	
CAT	GUA	
CAC	GUG	
CGA	GCU	Alanine
CGG	GCC	

Expression of the genetic code: transcription and translation

DNA represents a type of information that is vital to the shape and form of an organism. It contains instructions in a coded sequence of nucleotides, and this sequence interacts with the environment to produce form—the living organism with all of its complex structures and functions. The form of an organism is largely determined by protein. A large proportion of what we see when we observe the various parts of an organism is protein; for example, hair, muscle, and skin are made up largely of protein. Other chemical compounds that make up the human body, such as carbohydrates, fats, and more-complex chemicals, are either synthesized by catalytic proteins (enzymes) or are deposited at specific times and in specific tissues under the influence of proteins. For example, the black-brown skin pigment melanin is synthesized by enzymes and deposited in special skin cells called melanocytes. Genes exert their effect mainly by determining the structure and function of the many thousands of different proteins, which in turn determine the characteristics of an organism. Generally, it is true to say that each protein is coded for by one gene, bearing in mind that the production of some proteins requires the cooperation of several genes.

Proteins are polymeric molecules; that is, they are made up of chains of monomeric elements, as is DNA. In proteins, the monomers are amino acids. Organisms generally contain 20 different types of amino acids, and the distinguishing factors that make one protein different from another are its length and specific amino acid sequence, which are determined by the number and sequence of nucleotide pairs in DNA. In other words, there is a colinearity (i.e., parallel structure) between the polymer that is DNA and the polymer that is protein.

Hence, genetic information flows from DNA into protein. However, this is not a single-step process. First, the nucleotide sequence of DNA is copied into the nucleotide sequence of single-stranded RNA in a process called transcription. Transcription of any one gene takes place at the chromosomal location of that gene. Whereas the unit of replication is a whole chromosome, the transcriptional unit is a relatively short segment of the chromosome, the gene. The active transcription of a gene depends on the need for the activity of that particular gene in a specific tissue or at a given time.

How DNA directs protein synthesis

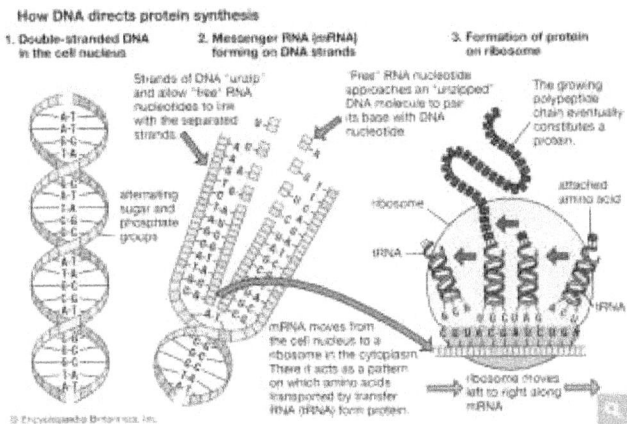

DNA in the cell nucleus carries a genetic code, which consists of sequences of adenine (A), thymine (T), guanine (G), and cytosine (C) (Figure 1). RNA, which contains uracil (U) instead of thymine, carries the code to protein-making sites in the cell. To make RNA, DNA pairs its bases with those of the "free" nucleotides (Figure 2). Messenger RNA (mRNA) then travels to the ribosomes in the cell cytoplasm, where protein synthesis occurs (Figure 3). The base triplets of transfer RNA (tRNA) pair with those of mRNA and at the same time deposit their amino acids on the growing protein chain. Finally, the synthesized protein is released to perform its task in the cell or elsewhere in the body.

The nucleotide sequence in RNA faithfully mirrors that of the DNA from which it was transcribed. The uracil in RNA has exactly the same hydrogen-bonding properties as thymine, so there are no changes at the information level. For most RNA molecules, the nucleotide sequence is converted into an amino acid sequence, a process called translation. In prokaryotes, translation begins during the transcription process, before the full RNA transcript is made. In eukaryotes, transcription finishes, and the RNA molecule passes from the nucleus into the cytoplasm, where translation takes place.

The genome of a type of virus called a retrovirus (of which the human immunodeficiency virus, or HIV, is an example) is composed of RNA instead of DNA. In a retrovirus, RNA is reverse transcribed into DNA, which can then integrate into the chromosomal DNA of the host cell that the retrovirus infects. The synthesis of DNA is catalyzed by the enzyme reverse transcriptase. The existence of reverse transcriptase shows that genetic information is capable of flowing from RNA to DNA in exceptional cases. Since it is believed that life arose in an RNA world, it is likely that the evolution of reverse transcriptase was an important step in the transition to the present DNA world.

Transcription

A gene is a functional region of a chromosome that is capable of making a transcript in response to appropriate regulatory signals. Therefore, a gene must not only be composed of the DNA sequence that is actually transcribed, but it must also include an adjacent regulatory, or control, region that is necessary for the transcript to be made in the correct developmental context.

The polymerization of ribonucleotides during transcription is catalyzed by the enzyme RNA polymerase. As with DNA replication, the two DNA strands must separate to expose the template. However, transcription differs from replication in that for any gene, only one of the DNA strands, the 3′ → 5′ strand, is actually used as a template. Synthesis of RNA is in the 5′ → 3′ direction, as with DNA. Hence, the growing point of the RNA chain is the 3′ end, and polymerization is continuous as the RNA polymerase moves along the transcribed region. The RNA strand is extruded from the transcription complex like a tail, which grows longer as the transcription process advances. Eventually, a full-length transcript of RNA is produced, and this detaches from the DNA. The process is repeated, and multiple RNA transcripts are produced from one gene.

Prokaryotes possess only one type of RNA polymerase, but in eukaryotes there are several different types. RNA polymerase I synthesizes ribosomal RNA (rRNA), and RNA polymerase III synthesizes transfer RNA (tRNA) and other small RNAs. The types of RNA transcribed by these two polymerases are never translated into protein. RNA polymerase II transcribes the major type of genes, those genes that code for proteins. Transcription of these genes is considered in detail below.

Transcription of protein-coding genes results in a type of RNA called messenger RNA (mRNA), so named because it carries a genetic message from the gene on a nuclear chromosome into the cytoplasm, where it is acted upon by the protein-synthesizing apparatus. The transcription machinery contains many items in addition to the RNA polymerase. The successful binding of the RNA polymerase to the DNA "upstream" of the transcribed sequence depends upon the cooperation of many additional proteinaceous transcription factors. The region of the gene upstream from the region to be transcribed contains specific DNA sequences that are essential for the binding of transcription factors and a region called the promoter, to which the RNA polymerase binds. These sequences must be a specific distance from the transcriptional start site for successful operation. Various short base sequences in this regulatory region physically bind specific transcription factors by virtue of a lock-and-key fit between the DNA and the protein. As might be expected, a protein binds with the centre of the DNA molecule, which contains the sequence specificity, and not with the outside of the molecule, which is merely a uniform repetition of sugar and phosphate groups.

In eukaryotes, a key segment is the TATA box, a TATA sequence approximately 30 nucleotides upstream from the transcription start site. If this sequence is changed or moved, the rate of transcription drops drastically. The TATA box is bound by a transcription factor called the TATA-binding protein, which, together with RNA polymerase II and numerous other transcription factors, assembles in a precise sequence around the TATA box, binding to each other and to the DNA. Together, RNA polymerase and the transcription factors constitute the transcription complex.

The RNA polymerase is directed by the transcription complex to begin transcription at the proper site. It then moves along the template, synthesizing mRNA as it goes. At some position past the coding region, the transcription process stops. Bacteria have well-characterized specific termination sequences; however, in eukaryotes, termination signals are less well understood, and the transcription process stops at variable positions past the end of the coding sequence. A short nucleotide sequence downstream from the coding region acts as a signal for the RNA to be cut at that position, and this becomes the 3' end of the new RNA strand. Subsequently, approximately 200 adenine nucleotides are added to the 3' end to form what is called a poly(A) tail, which is characteristic of all eukaryotic DNA. At the 5' end of the mRNA, a modified guanine nucleotide, called a cap, is added. Noncoding nucleotide sequences called introns are excised from the RNA at this stage in a process called intron splicing. Molecular complexes called spliceosomes, which are composed of proteins and RNA, have RNA sequences that are complementary to the junction between introns and adjacent coding regions called exons. The intron is twisted into a loop and excised, and the exons are linked together. The resulting capped, tailed, and intron-free molecule is now mature mRNA.

The genetic code

Hereditary information is contained in the nucleotide sequence of DNA in a kind of code. The coded information is copied faithfully into RNA and translated into chains of amino acids. Amino acid chains are folded into helices, zigzags, and other shapes and are sometimes associated with other amino acid chains. The specific amounts of amino acids in a protein and their sequence determine the protein's unique properties; for example, muscle protein and hair protein contain the same 20 amino acids, but the sequences of these amino acids in the two proteins are quite different. If the nucleotide sequence of mRNA is thought of as a written message, it can be said that this message is read by the translation apparatus in "words" of three nucleotides, starting at one end of the mRNA and proceeding along the length of the molecule. These three-letter words are called codons. Each codon stands for a specific amino acid, so if the message in mRNA is 900 nucleotides long, which corresponds to 300 codons, it will be translated into a chain of 300 amino acids.

Each of the three letters in a codon can be filled by any one of the four nucleotides; therefore, there are 43, or 64, possible codons. Each one of these 64 words in the codon dictionary has meaning. Most codons code for one of the 20 possible amino acids. Two amino acids, methionine and tryptophan, are each coded for by one codon only (AUG and UGG, respectively). The other 18 amino acids are coded for by two to six codons; for example, either of the codons UUU or UUC will cause the insertion of the amino acid phenylalanine into the growing amino acid chain. Three codons—UAG, UGA, and UAA— represent translation-termination signals and are called the stop codons. The first amino acid in an amino acid chain is methionine, encoded by an AUG codon. However, AUG codons are found throughout the coding sequence and are translated into methionines.

One of the surprising findings about the genetic codon dictionary is that, with a few exceptions, it is the same in all organisms. (One exception is mitochondrial DNA, which exhibits several differences from the standard genetic code and also between organisms.) The uniformity of the genetic code has been interpreted as an indication of the evolutionary relatedness of all organisms. For the purpose of genetic research, codon uniformity is convenient because any type of DNA can be translated in any organism.

Translation

The process of translation requires the interaction not only of large numbers of proteinaceous translational factors but also of specific membranes and organelles of the cell. In both prokaryotes and eukaryotes, translation takes place on cytoplasmic organelles called ribosomes. Ribosomes are aggregations of many different types of proteins and ribosomal RNA (rRNA). They can be thought of as cellular anvils on which the links of an amino acid chain are forged. A ribosome is a generic protein-making machine that can be recycled and used to synthesize many different types of proteins. A ribosome attaches to the 5' end of the mRNA, begins translation at the start codon AUG, and translates the message one codon at a time until a stop codon is reached. Any one mRNA is translated many times by several ribosomes along its length, each one at a different stage of translation. In eukaryotes, ribosomes that produce proteins to be used in the same cell are not associated with membranes. However, proteins that must be exported to another location in the organism are synthesized on ribosomes located on the outside of flattened membranous chambers called the endoplasmic reticulum (ER). A completed amino acid chain is extruded into the inner cavity of the ER. Subsequently, the ER transports the proteins via small vesicles to another cytoplasmic organelle called the Golgi apparatus, which in turn buds off more vesicles that eventually fuse with the cell membrane. The protein is then released from the cell.

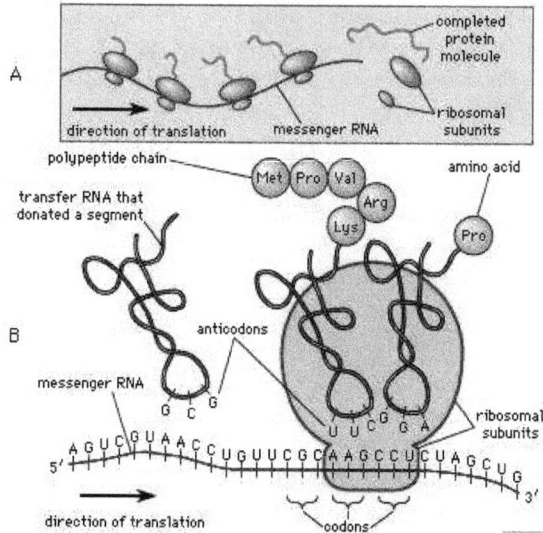

Synthesis of protein

Another crucial component of the translational process is transfer RNA (tRNA). The function of any one tRNA molecule is to bind to a designated amino acid and carry it to a ribosome, where the amino acid is added to the growing amino acid chain. Each amino acid has its own set of tRNA molecules that will bind only to that specific amino acid. A tRNA molecule is a single nucleotide chain with several helical regions and a loop containing three unpaired nucleotides, called an anticodon. The anticodon of any one tRNA fits perfectly into the mRNA codon that codes for the amino acid attached to that tRNA; for example, the mRNA codon UUU, which codes for the amino acid phenylalanine, will be bound by the anticodon AAA. Thus, any mRNA codon that happens to be on the ribosome at any one time will solicit the binding only of the tRNA with the appropriate anticodon, which will align the correct amino acid for addition to the chain. A tRNA molecule and its attached amino acid must bind to the ribosome as well as to the codon during this amino acid chain-elongation process. A ribosome has two tRNA binding sites; at the first site, one tRNA attaches to the amino acid chain, and at the second site, another tRNA carrying the next amino acid is attached. After attachment, the first tRNA departs and recycles, whereas the second tRNA is now left holding the amino acid chain. At this time the ribosome moves to the next codon, and the whole process is successively repeated along the length of the mRNA until a stop codon is reached, at which time the completed amino acid chain is released from the ribosome.

The amino acid chain then spontaneously folds to generate the three-dimensional shape necessary for its function. Each amino acid has its own special shape and pattern of electrical charges on its surface, and ultimately these are what determine the overall shape of the protein. The protein's shape is stabilized by weak bonds that form between different parts of the chain. In some proteins, strong covalent bridges are formed between two cysteines at different sites in the chain. If the protein is composed of two or more amino acid chains, these also associate spontaneously and take on their most stable three-dimensional shape. For enzymes, shape determines the ability to bind to its specific substrate (i.e., the substance on which an enzyme acts). For structural proteins, the amino acid sequence determines whether it will be a filament, a sheet, a globule, or another shape.

Gene mutation

Given the complexity of DNA and the vast number of cell divisions that take place within the lifetime of a multicellular organism, copying errors are likely to occur. If unrepaired, such errors will change the sequence of the DNA bases and alter the genetic code. Mutation is the random process whereby genes change from one allelic form to another. Scientists who study mutation use the most common genotype found in natural populations, called the wild type, as the standard against which to compare a mutant allele. Mutation can occur in two directions; mutation from wild type to mutant is called a forward mutation, and mutation from mutant to wild type is called a back mutation or reversion.

Mechanisms of mutation

Mutations arise from changes to the DNA of a gene. These changes can be quite small, affecting only one nucleotide pair, or they can be relatively large, affecting hundreds or thousands of nucleotides. Mutations in which one base is changed are called point mutations—for example, substitution of the nucleotide pair AT by GC, CG, or TA. Base substitutions can have different consequences at the protein level. Some base substitutions are "silent," meaning that they result in a new codon that codes for the same amino acid as the wild type codon at that position or a codon that codes for a different amino acid that happens to have the same properties as those in the wild type. Substitutions that result in a functionally different amino acid are called "missense" mutations; these can lead to alteration or loss of protein function. A more severe type of base substitution, called a "nonsense" mutation, results in a stop codon in a position where there was not one before, which causes the premature termination of protein synthesis and, more than likely, a complete loss of function in the finished protein.

Another type of point mutation that can lead to drastic loss of function is a frameshift mutation, the addition or deletion of one or more DNA bases. In a protein-coding gene, the sequence of codons starting with AUG and ending with a termination codon is called the reading frame. If a nucleotide pair is added to or subtracted from this sequence, the reading frame from that point will be shifted by one nucleotide pair, and all of the codons downstream will be altered. The result will be a protein whose first section (before the mutational site) is that of the wild type amino acid sequence, followed by a tail of functionally meaningless amino acids. Large deletions of many codons will not only remove amino acids from a protein but may also result in a frameshift mutation if the number of nucleotides deleted is not a multiple of three. Likewise, an insertion of a block of nucleotides will add amino acids to a protein and perhaps also have a frameshift effect.

A number of human diseases are caused by the expansion of a trinucleotide pair repeat. For example, fragile-X syndrome, the most common type of inherited mental retardation in humans, is caused by the repetition of up to 1,000 copies of a CGG repeat in a gene on the X chromosome.

The impact of a mutation depends upon the type of cell involved. In a haploid cell, any mutant allele will most likely be expressed in the phenotype of that cell. In a diploid cell, a dominant mutation will be expressed over the wild type allele, but a recessive mutation will remain masked by the wild type. If recessive mutations occur in both members of one gene pair in the same cell, the mutant phenotype will be expressed. Mutations in germinal cells (i.e., reproductive cells) may be passed on to successive generations. However, mutations in somatic (body) cells will exert their effect only on that individual and will not be passed on to progeny.

The impact of an expressed somatic mutation depends upon which gene has been mutated. In most cases, the somatic cell with the mutation will die, an event that is generally of little consequence in a multicellular organism. However, mutations in a special class of genes called proto-oncogenes can cause uncontrolled division of that cell, resulting in a group of cells that constitutes a cancerous tumour.

Mutations can affect gene function in several different ways. First, the structure and function of the protein coded by that gene can be affected. For example, enzymes are particularly susceptible to mutations that affect the amino acid sequence at their active site (i.e., the region that allows the enzyme to bind with its specific substrate). This may lead to enzyme inactivity; a protein is made, but it has no enzymatic function. Second, some nonsense or frameshift mutations can lead to the complete absence of a protein. Third, changes to the promoter region of the gene can result in gene malfunction by interfering with transcription. In this situation, protein production is either inhibited or it occurs at an inappropriate time because of alterations somewhere in the regulatory region. Fourth, mutations within introns that affect the specific nucleotide sequences that direct intron splicing may result in an mRNA that still contains an intron. When translated, this extra RNA will almost certainly be meaningless at the protein level, and its extra length will lead to a functionless protein. Any mutation that results in a lack of function for a particular gene is called a "null" mutation. Less-severe mutations are called "leaky" mutations because some normal function still "leaks through" into the phenotype.

Most mutations occur spontaneously and have no known cause. The synthesis of DNA is a cooperative venture of many different interacting cellular components, and occasionally mistakes occur that result in mutations. Like many chemical structures, the bases of DNA are able to exist in several conformations called isomers. The keto form of a DNA base is the normal form that gives the molecule its standard base-pairing properties. However, the keto form occasionally changes spontaneously to the enol form, which has different base-pairing properties. For example, the keto form of cytosine pairs with guanine (its normal pairing partner), but the enol form of cytosine pairs with adenine. During DNA replication, this adenine base will act as the template for thymine in the newly synthesized strand. Therefore, a CG base pair will have mutated to a TA base pair. If this change results in a functionally different amino acid, then a missense mutation may result. Another spontaneous event that can lead to mutation is depurination, the complete loss of a purine base (adenine or guanine) at some location in the DNA. The resulting gap can be filled by any base during subsequent replications.

Researchers have demonstrated that ionizing radiation, some chemicals, and certain viruses are capable of acting as mutagens—agents that can increase the rate at which mutations occur. Some mutagens have been implicated as a cause of cancer. For example, ultraviolet (UV) radiation from the sun is known to cause skin cancer, and cigarette smoke is a primary cause of lung cancer.

Repair of mutation

A variety of mechanisms exists for repairing copying errors caused by DNA damage. One of the best-studied systems is the repair mechanism for damage caused by ultraviolet radiation. Ultraviolet radiation joins adjacent thymines, creating thymine dimers, which, if not repaired, may cause mutations. Special repair enzymes either cut the bond between the thymines or excise the bonded dimer and replace it with two single thymines. If both of these repair methods fail, a third method allows the DNA replication process to bypass the dimer; however, it is this bypass system that causes most mutations because bases are then inserted at random opposite the thymine dimer. Xeroderma pigmentosum, a severe hereditary disease of humans, is caused by a mutation in a gene coding for one of the thymine dimer repair enzymes. Individuals with this disease are highly susceptible to skin cancer.

Reverse mutation from the aberrant state of a gene back to its normal, or wild type, state can result in a number of possible molecular changes at the protein level. True reversion is the reversal of the original nucleotide change. However, phenotypic reversion can result from changes that restore a different amino acid with properties identical to the original. Second-site changes within a protein can also restore normal function. For example, an amino acid change at a site different from that altered by the original mutation can sometimes interact with the amino acid at the first mutant site to restore a normal protein shape. Also, second-site mutations at other genes can act as suppressors, restoring wild type function. For example, mutations in the anticodon region of a tRNA gene can result in a tRNA that sometimes inserts an amino acid at an erroneous stop codon; if the original mutation is caused by a stop codon, which arrests translation at that point, then a tRNA anticodon change can insert an amino acid and allow translation to continue normally to the end of the mRNA. Alternatively, some mutations at separate genes open up a new biochemical pathway that circumvents the block of function caused by the original mutation.

Regulation of gene expression

Not all genes in a cell are active in protein production at any given time. Gene action can be switched on or off in response to the cell's stage of development and external environment. In multicellular organisms, different kinds of cells express different parts of the genome. In other words, a skin cell and a muscle cell contain exactly the same genes, but the differences in structure and function of these cells result from the selective expression and repression of certain genes.

In prokaryotes and eukaryotes, most gene-control systems are positive, meaning that a gene will not be transcribed unless it is activated by a regulatory protein. However, some bacterial genes show negative control. In this case the gene is transcribed continuously unless it is switched off by a regulatory protein. An example of negative control in prokaryotes involves three adjacent genes used in the metabolism of the sugar lactose by E. coli. The part of the chromosome containing the genes concerned is divided into two regions, one that includes the structural genes (i.e., those genes that together code for protein structure) and another that is a regulatory region. This overall unit is called an operon. If lactose is not present in a cell, transcription of the genes that code for the lactose-processing enzymes—β-galactosidase, permease, and transacetylase—is turned off. This is achieved by a protein called the lac repressor, which is produced by the repressor gene and binds to a region of the operon called the operator. Such binding prevents RNA polymerase, which initially binds at the adjacent promoter, from moving into the coding region. If lactose enters the cell, it binds to the lac repressor and induces a change of shape in the repressor so that it can no longer bind to the DNA at the operator. Consequently, the RNA polymerase is able to travel from the promoter down the three adjacent protein-coding regions, making one continuous transcript. This three-gene transcript is subsequently translated into three separate proteins.

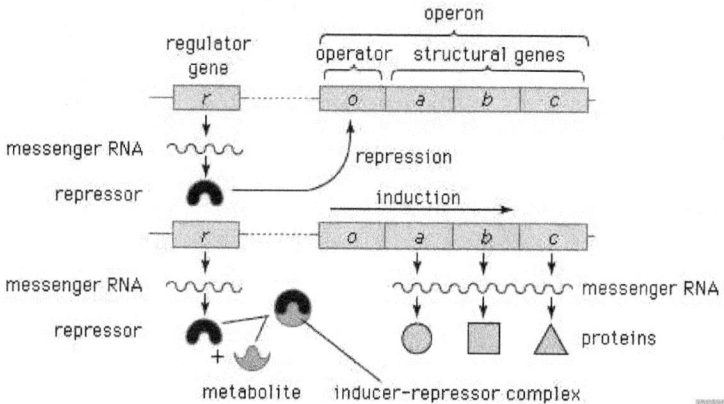

Model of the operon and its relation to the regulator gene.

Although the operon model has proved a useful model of gene regulation in bacteria, different regulatory mechanisms are employed in eukaryotes. First, there are no operons in eukaryotes, and each gene is regulated independently. Furthermore, the series of events associated with gene expression in higher organisms is much more complex than in prokaryotes and involves multiple levels of regulation.

In order for a gene to produce a functional protein, a complex series of steps must occur. Some type of signal must initiate the transcription of the appropriate region along the DNA, and, finally, an active protein must be made and sent to the appropriate location to perform its specific task. Regulation can be exerted at many different places along this pathway.

The fundamental level of control is the rate of transcription. Transcription itself is also a complex process with many different components, and each one is a potential point of control. Regulatory proteins called activators or enhancers are needed for the transcription of genes at a specific time or in a certain cell. Thus, control is positive (not negative as in the lac operon) in that these proteins are necessary for the promotion of transcription. Activators bind to specific regions of the DNA in the upstream regulatory region, some very distant from the binding of the initiation complex.

Following the transcription of DNA into RNA, a process of editing and splicing takes place in which noncoding nucleotide sequences called introns are excised from the primary transcript, resulting in functional mRNA. For most genes this is a routine step in the production of mRNA, but in some genes there are alternative ways to splice the primary transcript, resulting in different mRNAs, which in turn result in different proteins.

Some genes are controlled at the translational and post-translational levels. One type of translational control is the storage of uncapped mRNA to meet future demands for protein synthesis. In other cases, control is exerted through the stability or instability of mRNA. The rate of translation of some mRNAs can also be regulated. Post-translationally, certain proteins (e.g., insulin) are synthesized in an inactive form and must be chemically modified to become active. Other proteins are targeted to specific locations inside the cell (e.g., mitochondria) by means of highly specific amino acid sequences at their ends, called leader sequences; when the protein reaches its correct site, the leader segment is cut off, and the protein begins to function. Post-translational control is also exerted through mRNA and protein degradation.

Repetitive DNA

One major difference between the genomes of prokaryotes and eukaryotes is that most eukaryotes contain repetitive DNA, with the repeats either clustered or spread out between the unique genes. There are several categories of repetitive DNA: (1) single copy DNA, which contains the structural genes (protein-coding sequences), (2) families of DNA, in which one gene somehow copies itself, and the repeats are located in small clusters (tandem repeats) or spread throughout the genome (dispersed repeats), and (3) satellite DNA, which contains short nucleotide sequences repeated as many as thousands of times. Such repeats are often found clustered in tandem near the centromeres (i.e., the attachment points for the nuclear spindle fibres that move chromosomes during cell division).

Microsatellite DNA is composed of tandem repeats of two nucleotide pairs that are dispersed throughout the genome. Minisatellite DNA, sometimes called variable number tandem repeats (VNTRs), is composed of blocks of longer repeats also dispersed throughout the genome. There is no known function for satellite DNA, nor is it known how the repeats are created. There is a special class of relatively large DNA elements called transposons, which can make replicas of themselves that "jump" to different locations in the genome; most transposons eventually become inactive and no longer move, but, nevertheless, their presence contributes to repetitive DNA.

Secretory vesicles

The release of proteins or other molecules from a secretory vesicle is most often stimulated by a nervous or hormonal signal. For example, a nerve cell impulse triggers the fusion of secretory vesicles to the membrane at the nerve terminal, where the vesicles release neurotransmitters into the synaptic cleft (the gap between nerve endings). The action is one of exocytosis: the vesicle and the cell membrane fuse, allowing the proteins and glycoproteins in the vesicle to be released to the cell exterior.

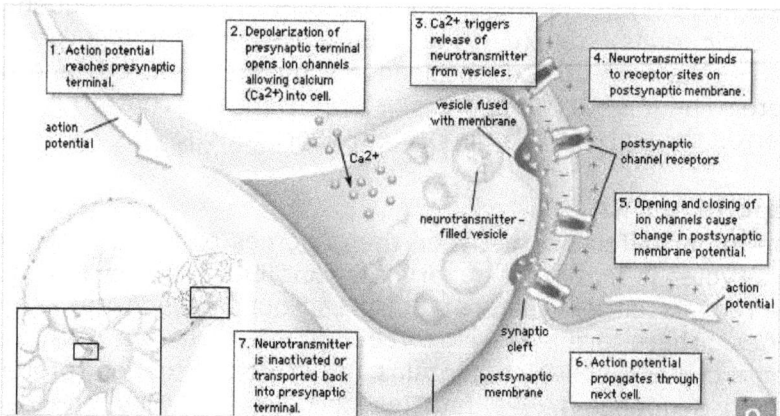

Chemical transmission of a nerve impulse at the synapseThe arrival of the nerve impulse at the presynaptic terminal stimulates the release of neurotransmitter into the synaptic gap. The binding of the neurotransmitter to receptors on the postsynaptic membrane stimulates the regeneration of the action potential in the postsynaptic neuron.

As secretory vesicles fuse with the cell membrane, the area of the cell membrane increases. Normal size is regained by the reuptake of membrane components through endocytosis. Regions bud in from the cell membrane and then fuse with internal membranes to effect recycling.

Sorting of products by chemical receptors

Not all proteins synthesized on the ER are destined for export. Many, such as the hydrolases in lysosomes, remain inside the cell; others become anchored in the membrane of internal organelles or in the cell membrane. It is presumed that each protein has some type of marker that fits a specific location in the cell.

Proteins synthesized on free ribosomes have segments that bind to specific receptors on the outer membrane of mitochondria, chloroplasts, or peroxisomes, allowing these proteins to be taken up only by these organelles. In the case of proteins synthesized in the RER, both the hydrolases destined for lysosomes and the secretory proteins are found initially in the same portion of the ER lumen. Studies have shown that these can be distinguished on the basis of their carbohydrate residues. The carbohydrate residues of lysosomal enzymes become modified in the cis-Golgi by the addition of certain phosphate groups. This critical modification allows the enzymes to bind to specific receptors on the membrane of the Golgi, which then directs them into vesicles leading to a lysosome rather than a secretory vesicle. In the lysosomes, proton pumps create an acidic environment that causes the release of the lysosomal enzyme from the membrane-bound receptors. Much of this sorting activity is mediated by coated vesicles containing the same fibrous outer protein, clathrin, used in endocytosis. These sorting vesicles also contain associated smaller proteins.

The Nucleus

The nucleus is the information centre of the cell and is surrounded by a nuclear membrane in all eukaryotic organisms. It is separated from the cytoplasm by the nuclear envelope, and it houses the double-stranded, spiral-shaped deoxyribonucleic acid (DNA) molecules, which contain the genetic information necessary for the cell to retain its unique character as it grows and divides.

The presence of a nucleus distinguishes the eukaryotic cells of multicellular organisms from the prokaryotic, one-celled organisms such as bacteria. In contrast to the higher organisms, prokaryotes do not have nuclei, so their DNA is maintained in the same compartment as their other cellular components.

The primary function of the nucleus is the expression of selected subsets of the genetic information encoded in the DNA double helix. Each subset of a DNA chain, called a gene, codes for the construction of a specific protein out of a chain of amino acids. Information in DNA is not decoded directly into proteins, however. First it is transcribed, or copied, into a range of messenger ribonucleic acid (mRNA) molecules, each of which encodes the information for one protein (or more than one protein in bacteria). The mRNA molecules are then transported through the nuclear envelope into the cytoplasm, where they are translated, serving as templates for the synthesis of specific proteins.

The nucleus must not only synthesize the mRNA for many thousands of proteins, but it must also regulate the amounts synthesized and supplied to the cytoplasm. Furthermore, the amounts of each type of mRNA supplied to the cytoplasm must be regulated differently in each type of cell. In addition to mRNA, the nucleus synthesizes and exports other classes of RNA involved in the mechanisms of protein synthesis.

Structural organization of the nucleus
DNA packaging

The nucleus of the average human cell is only 6 micrometres (6 × 10−6 metre) in diameter, yet it contains about 1.8 metres of DNA. This is distributed among 46 chromosomes, each consisting of a single DNA molecule about 40 mm (1.5 inches) long. The extraordinary packaging problem this poses can be envisaged by a scale model enlarged a million times. On this scale a DNA molecule would be a thin string 2 mm thick, and the average chromosome would contain 40 km (25 miles) of DNA. With a diameter of only 6 metres, the nucleus would contain 1,800 km (1,118 miles) of DNA.

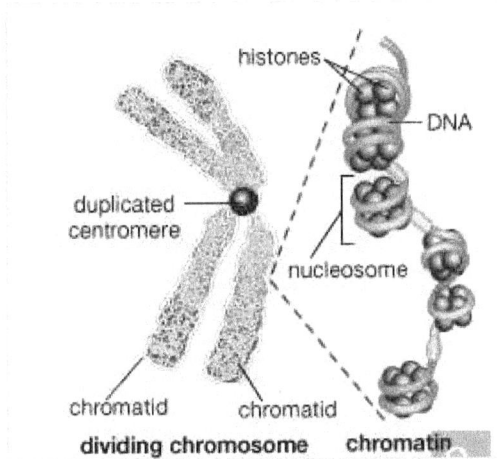

dividing chromosome   chromatin

During the first stages of cell division, the recognizable double-stranded chromosome is formed by two tightly coiled DNA strands (chromatids) joined at a point called the centromere. During the middle stage of cell division, the centromere duplicates, and the chromatid pair separates. Following cell division, the separated chromatids uncoil; the loosely coiled DNA, wrapped around its associated proteins (histones) to form beaded structures called nucleosomes, is termed chromatin.

These contents must be organized in such a way that they can be copied into RNA accurately and selectively. DNA is not simply crammed or wound into the nucleus like a ball of string; rather, it is organized, by molecular interaction with specific nuclear proteins, into a precisely packaged structure. This combination of DNA with proteins creates a dense, compact fibre called chromatin. An extreme example of the ordered folding and compaction that chromatin can undergo is seen during cell division, when the chromatin of each chromosome condenses and is divided between two daughter cells

Nucleosomes: the subunits of chromatin

The compaction of DNA is achieved by winding it around a series of small proteins called histones. Histones are composed of positively charged amino acids that bind tightly to and neutralize the negative charges of DNA. There are five classes of histone. Four of them, called H2A, H2B, H3, and H4, contribute two molecules each to form an octamer, an eight-part core around which two turns of DNA are wrapped. The resulting beadlike structure is called the nucleosome. The DNA enters and leaves a series of nucleosomes, linking them like beads along a string in lengths that vary between species of organism or even between different types of cell within a species. A string of nucleosomes is then coiled into a solenoid configuration by the fifth histone, called H1. One molecule of H1 binds to the site at which DNA enters and leaves each nucleosome, and a chain of H1 molecules coils the string of nucleosomes into the solenoid structure of the chromatin fibre.

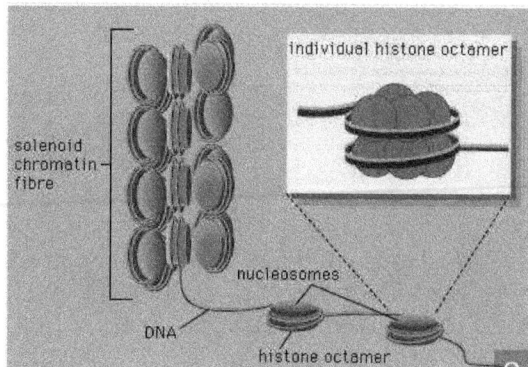

DNA wrapped around clusters of histone proteins to form nucleosomes, which can coil to form solenoids.

Nucleosomes not only neutralize the charges of DNA, but they have other consequences. First, they are an efficient means of packaging. DNA becomes compacted by a factor of six when wound into nucleosomes and by a factor of about 40 when the nucleosomes are coiled into a solenoid chromatin fibre. The winding into nucleosomes also allows some inactive DNA to be folded away in inaccessible conformations, a process that contributes to the selectivity of gene expression.

Organization of chromatin fibre

Several studies indicate that chromatin is organized into a series of large radial loops anchored to specific scaffold proteins. Each loop consists of a chain of nucleosomes and may be related to units of genetic organization. This radial arrangement of chromatin loops compacts DNA about a thousandfold. Further compaction is achieved by a coiling of the entire looped chromatin fibre into a dense structure called a chromatid, two of which form the chromosome. During cell division, this coiling produces a 10,000-fold compaction of DNA.

The nuclear envelope

The nuclear envelope is a double membrane composed of an outer and an inner phospholipid bilayer. The thin space between the two layers connects with the lumen of the rough endoplasmic reticulum (RER), and the outer layer is an extension of the outer face of the RER.

The inner surface of the nuclear envelope has a protein lining called the nuclear lamina, which binds to chromatin and other contents of the nucleus. The entire envelope is perforated by numerous nuclear pores. These transport routes are fully permeable to small molecules up to the size of the smallest proteins, but they form a selective barrier against movement of larger molecules. Each pore is surrounded by an elaborate protein structure called the nuclear pore complex, which selects molecules for entrance into the nucleus. Entering the nucleus through the pores are the nucleotide building blocks of DNA and RNA, as well as adenosine triphosphate, which provides the energy for synthesizing genetic material. Histones and other large proteins must also pass through the pores. These molecules have special amino acid sequences on their surface that signal admittance by the nuclear pore complexes. The complexes also regulate the export from the nucleus of RNA and subunits of ribosomes.

DNA in prokaryotes is also organized in loops and is bound to small proteins resembling histones, but these structures are not enclosed by a nuclear membrane.

Genetic organization of the nucleus
The structure of DNA

Several features are common to the genetic structure of most organisms. First is the double-stranded DNA. Each strand of this molecule is a series of nucleotides, and each nucleotide is composed of a sugar-phosphate compound attached to one of four nitrogen-containing bases. The sugar-phosphate compounds link together to form the backbone of the strand. Each of the bases strung along the backbone is chemically attracted to a corresponding base on the parallel strand of the DNA molecule. This base pairing joins the two strands of the molecule much as rungs join the two sides of a ladder, and the chemical bonding of the base pairs twists the doubled strands into a spiral, or helical, shape.

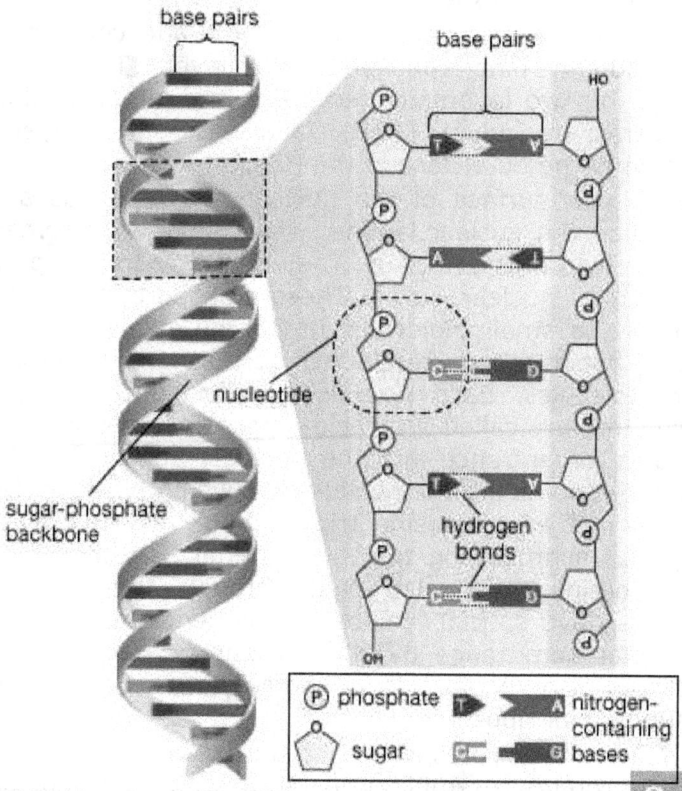

DNA molecule

    The four nucleotide bases are adenine, cytosine, guanine, and thymine. DNA is composed of millions of these bases strung in an apparently limitless variety of sequences. It is in the sequence of bases that the genetic information is contained, each sequence determining the sequence of amino acids to be connected into proteins. A nucleotide sequence sufficient to encode one protein is called a gene. Genes are interspersed along the DNA molecule with other sequences that do not encode proteins. Some of these so-called untranslated regions regulate the activity of the adjacent genes, for example, by marking the points at which enzymes begin and cease transcribing DNA into RNA.

Rearrangement and modification of DNA

Rearrangements and modifications of the nucleotide sequences in DNA are exceptions to the rules of genetic expression and sometimes cause significant changes in the structure and function of cells. Different cells of the body owe their specialized structures and functions to different genes. This does not mean that the set of genetic information varies among the cells of the body. Indeed, for each cell the entire DNA content of the chromosomes is usually duplicated exactly from generation to generation, and, in general, the genetic content and arrangement is strikingly similar among different cell types of the same organism. As a result, the differentiation of cells can occur without the loss or irreversible inactivation of unnecessary genes, an observation that is reinforced by the presence of specific genes in a range of adult tissues. For example, normal copies of the genes encoding hemoglobin are present in the same numbers in red blood cells, which make hemoglobin, as in a range of other types of cells, which do not.

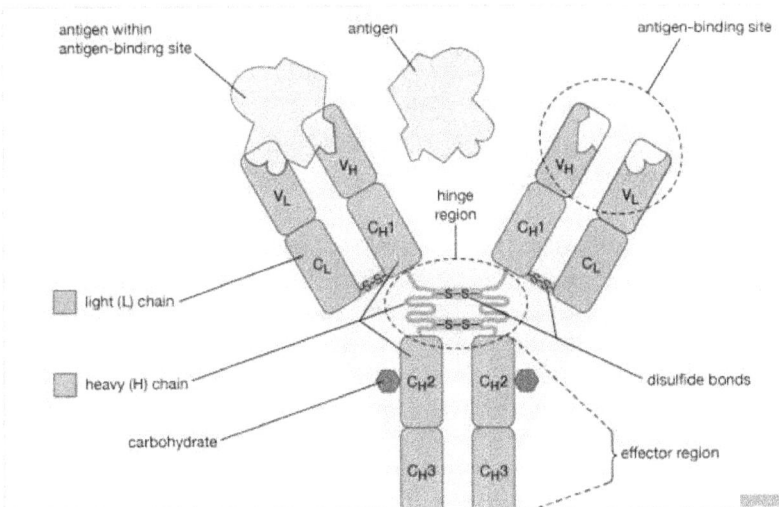

The structure of an antibody molecule represents the dramatic rearrangements of DNA that occur in the immune systems of mammals. Each antibody contains a light chain and a heavy chain that are encoded by different segments of DNA. These segments are subject to considerable variation and are thus able to produce many different antibodies.

Despite the general uniformity of genetic content in all the cells of an organism, studies have shown a few clear examples in some organisms of programmed, reversible change in the DNA of developing tissues. One of the most dramatic rearrangements of DNA occurs in the immune systems of mammals. The body's defense against invasion by foreign organisms involves the synthesis of a vast range of antibodies by lymphocytes (a type of white blood cell). Antibodies are proteins that bind to specific invading molecules or organisms and either inactivate them or signal their destruction. The binding sites on each antibody molecule are formed by one light and one heavy amino acid chain, which are encoded by different segments of the DNA in the lymphocyte nucleus. These DNA segments undergo considerable rearrangements, resulting in the synthesis of a great variety of antibodies. Some invasive organisms, such as trypanosome parasites, which cause sleeping sickness, go to great lengths to rearrange their own DNA to evade the versatility of their hosts' antibody production. The parasites are covered by a thick coat of glycoprotein (a protein with sugars attached). Given time, host organisms can overcome infection by producing antibodies to the parasites' glycoprotein coat, but this reaction is anticipated and evaded by the selective rearrangement of the trypanosomes' DNA encoding the glycoprotein, thus constantly changing the surface presented to the hosts' immune system

Careful comparisons of gene structure have also revealed epigenetic modifications, heritable changes that occur on the sugar-phosphate side of bases in the DNA and thus do not cause rearrangements in the DNA sequence itself. An example of an epigenetic modification involves the addition of a methyl group to cytosine bases. This appears to cause the inactivation of genes that do not need to be expressed in a particular type of cell. An important feature of the methylation of cytosine lies in its ability to be copied, so that methyl groups in a dividing cell's DNA will result in methyl groups in the same positions in the DNA of both daughter cells.

Genetic expression through RNA
The transcription of the genetic code from DNA to RNA, and the translation of that code from RNA into protein, exerts the greatest influence on the modulation of genetic information. The process of genetic expression takes place over several

stages, and at each stage is the potential for further differentiation of cell types.

As explained above, genetic information is encoded in the sequences of the four nucleotide bases making up a DNA molecule. One of the two DNA strands is transcribed exactly into messenger RNA (mRNA), with the exception that the thymine base of DNA is replaced by uracil. RNA also contains a slightly different sugar component (ribose) from that of DNA (deoxyribose) in its connecting sugar-phosphate chain. Unlike DNA, which is stable throughout the cell's life and of which individual strands are even passed on to many cell generations, RNA is unstable. It is continuously broken down and replaced, enabling the cell to change its patterns of protein synthesis.

Apart from mRNA, which encodes proteins, other classes of RNA are made by the nucleus. These include ribosomal RNA (rRNA), which forms part of the ribosomes and is exported to the cytoplasm to help translate the information in mRNA into proteins. Ribosomal RNA is synthesized in a specialized region of the nucleus called the nucleolus, which appears as a dense area within the nucleus and contains the genes that encode rRNA. This is also the site of assembly of ribosome subunits from rRNA and ribosomal proteins. Ribosomal proteins are synthesized in the cytoplasm and transported to the nucleus for subassembly in the nucleolus. The subunits are then returned to the cytoplasm for final assembly. Another class of RNA synthesized in the nucleus is transfer RNA (tRNA), which serves as an adaptor, matching individual amino acids to the nucleotide triplets of mRNA during protein synthesis.

RNA synthesis

The synthesis of RNA is performed by enzymes called RNA polymerases. In higher organisms there are three main RNA polymerases, designated I, II, and III (or sometimes A, B, and C). Each is a complex protein consisting of many subunits. RNA polymerase I synthesizes three of the four types of rRNA (called 18S, 28S, and 5.8S RNA); therefore it is active in the nucleolus, where the genes encoding these rRNA molecules reside. RNA polymerase II synthesizes mRNA, though its initial products are not mature RNA but larger precursors, called heterogeneous nuclear RNA, which are completed later. The products of RNA polymerase III include tRNA and the fourth RNA component of the ribosome, called 5S RNA.

All three polymerases start RNA synthesis at specific sites on DNA and proceed along the molecule, linking selected nucleotides sequentially until they come to the end of the gene and terminate the growing chain of RNA. Energy for RNA synthesis comes from high-energy phosphate linkages contained in the nucleotide precursors of RNA. Each unit of the final RNA product is essentially a sugar, a base, and one phosphate, but the building material consists of a sugar, a base, and three phosphates. During synthesis two phosphates are cleaved and discarded for each nucleotide that is incorporated into RNA. The energy released from the phosphate bonds is used to link the nucleotides. The crucial feature of RNA synthesis is that the sequence of nucleotides joined into a growing RNA chain is specified by the sequence of nucleotides in the DNA template: each adenine in DNA specifies uracil in RNA, each cytosine specifies guanine, each guanine specifies cytosine, and each thymine in DNA specifies adenine. In this way the information encoded in each gene is transcribed into RNA for translation by the protein-synthesizing machinery of the cytoplasm.

In addition to specifying the sequence of amino acids to be polymerized into proteins, the nucleotide sequence of DNA contains supplementary information. For example, short sequences of nucleotides determine the initiation site for each RNA polymerase, specifying where and when RNA synthesis should occur. In the case of RNA polymerases I and II, the sequences specifying initiation sites lie just ahead of the genes.

In contrast, the equivalent information for RNA polymerase III lies within the gene—that is, within the region of DNA to be copied into RNA. The initiation site on a segment of DNA is called a promoter. The promoters of different genes have some nucleotide sequences in common, but they differ in others. The differences in sequence are recognized by specific proteins called transcription factors, which are necessary for the expression of particular types of genes. The specificity of transcription factors contributes to differences in the gene expression of different types of cells.

Processing of mRNA
During and after synthesis, mRNA precursors undergo a complex series of changes before the mature molecules are released from the nucleus. First, a modified nucleotide is added to the start of the RNA molecule by a reaction called capping.

This cap later binds to a ribosome in the cytoplasm. The synthesis of mRNA is not terminated simply by the RNA polymerase's detachment from DNA, but by chemical cleavage of the RNA chain. Many (but not all) types of mRNA have a simple polymer of adenosine residues added to their cleaved ends.

In addition to these modifications of the termini, startling discoveries in 1977 revealed that portions of newly synthesized RNA molecules are cut out and discarded. In many genes, the regions coding for proteins are interrupted by intervening sequences of nucleotides called introns. These introns must be excised from the RNA copy before it can be released from the nucleus as a functional mRNA. The number and size of introns within a gene vary greatly, from no introns at all to more than 50. The sum of the lengths of these intervening sequences is sometimes longer than the sum of the regions coding for proteins.

The removal of introns, called RNA splicing, appears to be mediated by small nuclear ribonucleoprotein particles (snRNP's). These particles have RNA sequences that are complementary to the junctions between introns and adjacent coding regions. By binding to the junction ends, an snRNP twists the intron into a loop. It then excises the loop and splices the coding regions.

Regulation of genetic expression

Although all the cell nuclei of an organism generally carry the same genes, there are conspicuous differences between the specialized cell types of the body. The source of these differences lies not so much in the occasional modification of DNA, as outlined above, but in the selective expression of DNA through RNA; in particular, it can be traced to processes regulating the amounts and activities of mRNA both during and after its synthesis in the nucleus.

Regulation of RNA synthesis

The first level of regulation is mediated by variations in chromatin structure. In order to be transcribed, a gene must be assembled into a structurally distinct form of active chromatin. A second level of regulation is achieved by varying the frequency with which a gene in the active conformation is transcribed into RNA by an RNA polymerase. There is evidence for regulation of RNA synthesis at both these levels—for example, in response to hormone induction. At both levels, protein factors are believed to perform the regulation—for example, by binding to special promoter DNA regions flanking the transcribed gene.

Regulation of RNA after synthesis

After synthesis, RNA molecules undergo selective processing, which results in the export of only a subpopulation of RNA molecules to the cytoplasm. Furthermore, the stability in the cytoplasm of a particular type of mRNA can be regulated. For example, the hormone prolactin increases synthesis of milk proteins in tissue by causing a twofold rise in the rate of mRNA synthesis; but it also causes a 17-fold rise in mRNA lifetime, so that in this case the main cause of increased protein synthesis is the prolonged availability of mRNA. Conversely, there is evidence for selective destabilization of some mRNA—such as histone mRNA, which is rapidly broken down when DNA replication is interrupted. Finally, there are many examples of selective regulation of the translation of mRNA into protein.

The Mitochondrion And The Chloroplast

Mitochondria and chloroplasts are the powerhouses of the cell. Mitochondria appear in both plant and animal cells as elongated cylindrical bodies, roughly one micrometre in length and closely packed in regions actively using metabolic energy. Mitochondria oxidize the products of cytoplasmic metabolism to generate adenosine triphosphate (ATP), the energy currency of the cell. Chloroplasts are the photosynthetic organelles in plants and some algae. They trap light energy and convert it partly into ATP but mainly into certain chemically reduced molecules that, together with ATP, are used in the first steps of carbohydrate production. Mitochondria and chloroplasts share a certain structural resemblance, and both have a somewhat independent existence within the cell, synthesizing some proteins from instructions supplied by their own DNA.

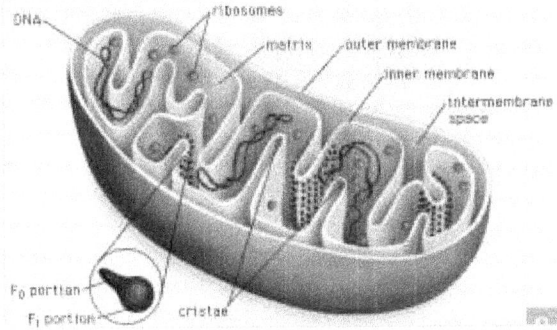

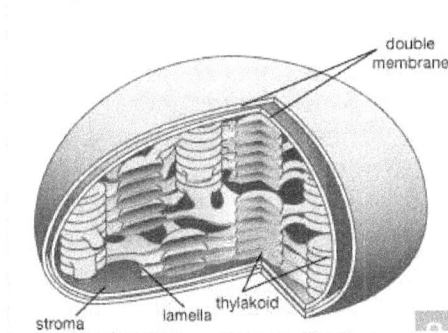

Mitochondrial and chloroplastic structure

Both organelles are bounded by an external membrane that serves as a barrier by blocking the passage of cytoplasmic proteins into the organelle. An inner membrane provides an additional barrier that is impermeable even to small ions such as protons. The membranes of both organelles have a lipid bilayer construction. Located between the inner and outer membranes is the intermembrane space.

In mitochondria the inner membrane is elaborately folded into structures called cristae that dramatically increase the surface area of the membrane. In contrast, the inner membrane of chloroplasts is relatively smooth. However, within this membrane is yet another series of folded membranes that form a set of flattened, disklike sacs called thylakoids. The space enclosed by the inner membrane is called the matrix in mitochondria and the stroma in chloroplasts. Both spaces are filled with a fluid containing a rich mixture of metabolic products, enzymes, and ions. Enclosed by the thylakoid membrane of the chloroplast is the thylakoid space. The extraordinary chemical capabilities of the two organelles lie in the cristae and the thylakoids. Both membranes are studded with enzymatic proteins either traversing the bilayer or dissolved within the bilayer. These proteins contribute to the production of energy by transporting material across the membranes and by serving as electron carriers in important oxidation-reduction reactions.

Metabolic functions

Crucial to the function of mitochondria and chloroplasts is the chemistry of the oxidation-reduction, or redox, reaction. This controlled burning of material comprises the transfer of electrons from one compound, called the donor, to another, called the acceptor. All compounds taking part in redox reactions are ranked in a descending scale according to their ability to act as electron donors. Those higher in the scale donate electrons to their fellows lower down, which have a lesser tendency to donate, but a correspondingly greater tendency to accept, electrons. Each acceptor in turn donates electrons to the next compound down the scale, forming a donor-acceptor chain extending from the greatest donating ability to the least.

At the top of the scale is hydrogen, the most abundant element in the universe. The nucleus of a hydrogen atom is composed of one positively charged proton; around the nucleus revolves one negatively charged electron. In the atmosphere two hydrogen atoms join to form a hydrogen molecule (H_2). In solution the two atoms pull apart, dissociating into their constituent protons and electrons. In the redox reaction the electrons are passed from one reactant to another. The donation of electrons is called oxidation, and the acceptance is called reduction—hence the descriptive term oxidation-reduction, indicating that one action never takes place without the other.

A hydrogen atom has a great tendency to transfer an electron to an acceptor. An oxygen atom, in contrast, has a great tendency to accept an electron. The burning of hydrogen by oxygen is, chemically, the transfer of an electron from each of two hydrogen atoms to oxygen, so that hydrogen is oxidized and oxygen reduced. The reaction is extremely exergonic; i.e., it liberates much free energy as heat. This is the reaction that takes place within mitochondria but is so controlled that the heat is liberated not at once but in a series of steps. The free energy, harnessed by the organelle, is coupled to the synthesis of ATP from adenosine diphosphate (ADP) and inorganic phosphate (Pi).

An analogy can be drawn between this controlled reaction and the flow of river water down a lock system. Without the locks, water flow would be rapid and uncontrolled, and no ship could safely ply the river. The locks force water to flow in small controlled steps conducive to safe navigation. But there is more to a lock system than this. The flow of water down the locks can also be harnessed to raise a ship from a lower to a higher level, with the water rather than the ship expending the energy. In mitochondria the burning of hydrogen is broken into a series of small indirect steps following the flow of electrons along a chain of donor-acceptors. Energy is funneled into the chemical bonding of ADP and Pi, raising the free energy of these two compounds to the high level of ATP.

The mitochondrion
Formation of the electron donors NADH and FADH2
Through a series of metabolic reactions carried out in the matrix, the mitochondrion converts products of the cell's initial metabolism of fats, amino acids, and sugars into the compound acetyl coenzyme A. The acetate portion of this compound is then oxidized in a chain reaction called the tricarboxylic acid cycle. At the end of this cycle the carbon atoms yield carbon dioxide and the hydrogen atoms are transferred to the cell's most important hydrogen acceptors, the coenzymes nicotinamide adenine dinucleotide (NAD+) and flavin adenine dinucleotide (FAD), yielding NADH and FADH2. It is the subsequent oxidation of these hydrogen acceptors that leads eventually to the production of ATP.

NADH and FADH2 are compounds of high electron-donating capacity. Were they to transfer their electrons directly to oxygen, the resulting combustion would release a lethal burst of heat energy. Instead, the energy is released in a series of electron donor-acceptor reactions carried out within the cristae of the mitochondrion by a number of proteins and coenzymes that make up the electron-transport, or respiratory, chain.

The electron-transport chain
The proteins of this chain are embedded in the cristae membrane, actually traversing the lipid bilayer and protruding from the inner and outer surfaces. The coenzymes are

dissolved in the lipid and diffuse through the membrane or across its surface. The proteins are arranged in three large complexes, each composed of a number of polypeptide chains. Each complex is, to continue the hydraulic analogy, a lock in the waterfall of the electron flow and the site at which energy from the overall redox reaction is tapped. The first complex, NADH dehydrogenase, accepts a pair of electrons from the primary electron donor NADH and is reduced in the process. It in turn donates these electrons to the coenzyme ubiquinone, a lipid-soluble molecule composed of a substituted benzene ring attached to a hydrocarbon tail. Ubiquinone, diffusing through the lipid of the cristae membrane, reaches the second large complex of the electron-transport chain, the b-c2 complex, which accepts the electrons, oxidizing ubiquinone and being itself reduced. (This complex can also accept electrons from the second primary electron donor, FADH2, a molecule below NADH in the electron-donating scale.) The b-c2 complex transfers the pair of electrons to cytochrome c, a small protein situated on the outer surface of the cristae membrane. From cytochrome c, electrons pass (four at a time) to the third large complex, cytochrome oxidase, which, in the final step of the chain, transfers the four electrons to two oxygen atoms and two protons, generating two water molecules.

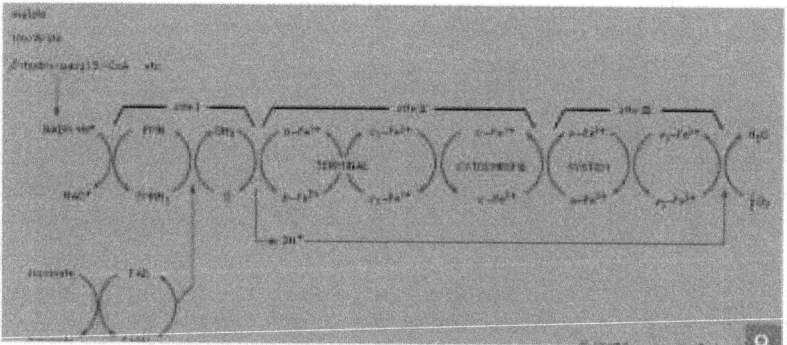

The electron-transport chain embedded in the inner membrane of a mitochondrion is made up of a series of electron donors and electron acceptors. The transport of electrons begins with the acceptance of electrons by NADH dehydrogenase from NADH. The electrons are then passed to ubiquinone (coenzyme Q; site I), which carries them to the b-c2 complex. The electrons are then transferred to cytochrome c (site II), to cytochrome oxidase (site III), and finally to oxygen.

The electron-transport chain embedded in the inner membrane of a mitochondrion is made up of a series of electron donors and electron acceptors. The transport of electrons begins with the acceptance of electrons by NADH dehydrogenase from NADH. The electrons are then passed to ubiquinone (coenzyme Q; site I), which carries them to the b-c2 complex. The electrons are then transferred to cytochrome c (site II), to cytochrome oxidase (site III), and finally to oxygen.

This transfer of electrons, from member to member of the electron-transport chain, provides energy for the synthesis of ATP through an indirect route. At the beginning of the electron-transport chain, NADH and FADH2 split hydrogen atoms into protons and electrons, transferring the electrons to the next protein complex and releasing the protons into the mitochondrial matrix. When each protein complex in turn transfers the electrons down the chain, it uses the energy released in this process to pump protons across the inner membrane into the intermembrane space. (For the dynamics of this pumping action, see above Transport across the membrane.) This transport of positively charged protons into the intermembrane space, opposite the negatively charged electrons in the matrix, creates an electrical potential that tends to draw the protons back across the membrane. A high concentration of protons outside the membrane also creates the conditions for their diffusion back into the matrix. However, as explained above, the inner membrane is extremely impermeable to protons. In order for the protons to flow back down the electrochemical gradient, they must traverse the membrane through transport molecules similar to the protein complexes of the electron-transfer chain. These molecules are the so-called F1F0ATPase, a complex protein that, transporting protons back into the matrix, uses the energy released to synthesize ATP. The protons then join the electrons and oxygen atoms to form water. (For further discussion of ATP production, see above Coupled chemical reactions.)

This complex chain of events, the basis of the cell's ability to derive ATP from metabolic oxidation, was conceived in its entirety by the British biochemist Peter Mitchell in 1961. The years following the announcement of his chemiosmotic theory saw its ample substantiation and revealed its profound implications for cell biology.

The chemiosmotic theory

The four postulates of the chemiosmotic theory, including examples of their experimental substantiation, are as follows:

(1) The inner mitochondrial membrane is impermeable to protons, hydroxide ions, and other cations and anions. This postulate was validated when it was shown that substances allowing protons to flow readily across mitochondrial membranes uncouple oxidative electron transport from ATP production.

(2) Transfer of electrons down the electron-transport chain brings about pumping of protons across the inner membrane, from matrix to intermembrane space. This was demonstrated in laboratory experiments that reconstituted the components of the electron-transport chain in artificial membrane vesicles. The stimulation of electron transport caused a measurable buildup of protons within the vesicle.

(3) The flow of protons down a built-up electrochemical gradient occurs through a proton-dependent ATPase, so that ATP is synthesized from ADP and Pi whenever protons move through the enzyme. This hypothesis was confirmed by the discovery of what came to be known as the F1F0ATPase. Shaped like a knob attached to the membrane by a narrow stalk, F1F0ATPase covers the inner surface of the cristae. Its stalk (the F0 portion) penetrates the lipid bilayer of the inner membrane and is capable of catalyzing the transport of protons. The knob (the F1 portion) is capable of synthesizing as well as splitting, or hydrolyzing, ATP. F1F0ATPase is therefore reversible, either using the energy of proton diffusion to combine ADP and Pi or using the energy of ATP hydrolysis to pump protons out of the matrix.

(4) The inner membrane of the mitochondrion possesses a complement of proteins that brings about the transport of essential metabolites. Numerous carrier systems have been demonstrated to transport into the mitochondrion the products of metabolism that are transformed into substrates for the electron-transport chain. Best known is the ATP-ADP exchange carrier of the inner membrane. Neither ATP nor ADP, being large charged molecules, can cross the membrane unaided, but ADP must enter and ATP must leave the mitochondrial matrix in order for ATP synthesis to continue. A single protein conducts the counter-transport of ATP against ADP, the energy released by the flow of ATP down its concentration gradient being

coupled to the pumping of ADP up its gradient and into the mitochondrion.

The chloroplast
Trapping of light

Light travels as packets of energy known as photons and is absorbed in this form by light-absorbing chlorophyll molecules embedded in the thylakoid membrane of the chloroplast. The chlorophyll molecules are grouped into antenna complexes, clusters of several hundred molecules that are anchored onto the thylakoid membrane by special proteins. Within each antenna complex is a specialized set of proteins and chlorophyll molecules that form a reaction centre. Photons absorbed by the other chlorophylls of the antenna are funneled into the reaction centre. The energy of the photon is absorbed by an electron of the reaction centre molecule in sufficient quantity to enable its acceptance by a nearby coenzyme, which cannot accept electrons at low energy levels. This coenzyme has a high electron-donor capability; it initiates the transfer of the electron down an electron-transport chain similar to that of the mitochondrion. Meanwhile, the loss of the negatively charged electron leaves a positively charged "hole" in the reaction centre chlorophyll molecule. This hole is filled by the enzymatic splitting of water into molecular oxygen, protons, and electrons and the transfer of an electron to the chlorophyll. The oxygen is released by the chloroplast, making its way out of the plant and into the atmosphere. The protons, in a process similar to that in the mitochondrion, are pumped through the thylakoid membrane and into the thylakoid space. Their facilitated diffusion back into the stroma through proteins embedded in the membrane powers the synthesis of ATP. This part of the photosynthetic process is called photosystem II.

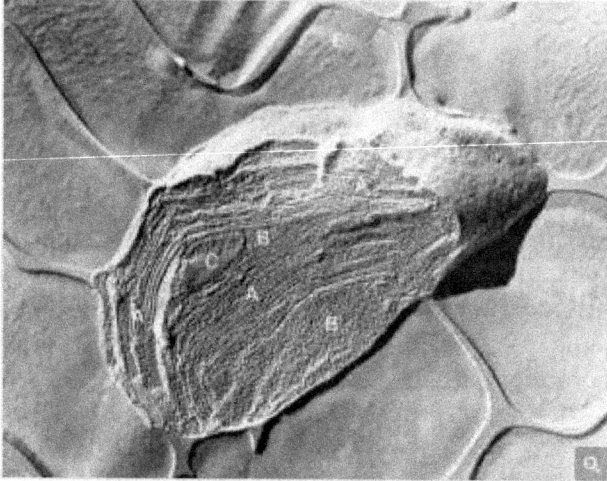

Electron micrograph of an isolated spinach chloroplast.
Courtesy of the Lawrence Berkeley National Laboratory,
University of California, Berkeley

At the end of the electron-transport chain in the thylakoid membrane is another reaction centre molecule. The electron is again energized by photons and then transported down another chain, which makes up photosystem I. This system uses the energy released in electron transfer to join a proton to nicotinamide adenine dinucleotide phosphate (NADP+), a phosphorylated derivative of NAD+, forming NADPH. NADPH is a high-energy electron donor that, with ATP, fuels the conversion of carbon dioxide into the carbohydrate foods of the plant cell.

The reaction of photosynthesis

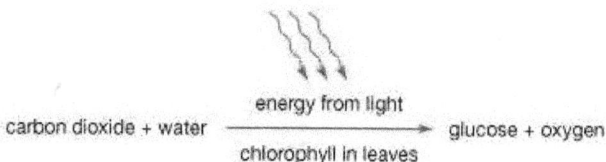

energy from light

carbon dioxide + water $\longrightarrow$ glucose + oxygen

chlorophyll in leaves

$$6CO_2 + 6H_2O \longrightarrow C_6H_{12}O_6 + 6O_2$$

In photosynthesis, plants consume carbon dioxide and water and produce glucose and oxygen. Energy for this process is provided by light, which is absorbed by pigments, primarily

76

chlorophyll. Chlorophyll is the pigment that gives plants their green colour.

Fixation of carbon dioxide

NADPH remains within the stroma of the chloroplast for use in the fixation of carbon dioxide (CO_2) during the Calvin cycle. In a complex cycle of chemical reactions, CO_2 is bound to a five-carbon ribulose biphosphate compound. The resulting six-carbon intermediate is then split into three-carbon phosphoglycerate. With energy supplied by the breakdown of NADPH and ATP, this compound is eventually formed into glyceraldehyde 3-phosphate, an important sugar intermediate of metabolism. One glyceraldehyde molecule is exported from the chloroplast, for further conversion in the cytoplasm, for every five that undergo an ATP-powered re-formation into the five-carbon ribulose biphosphate. In this way three molecules of CO_2 yield one molecule of glyceraldehyde 3-phosphate, while the entire fixation cycle hydrolyzes nine molecules of ATP and oxidizes six molecules of NADPH.

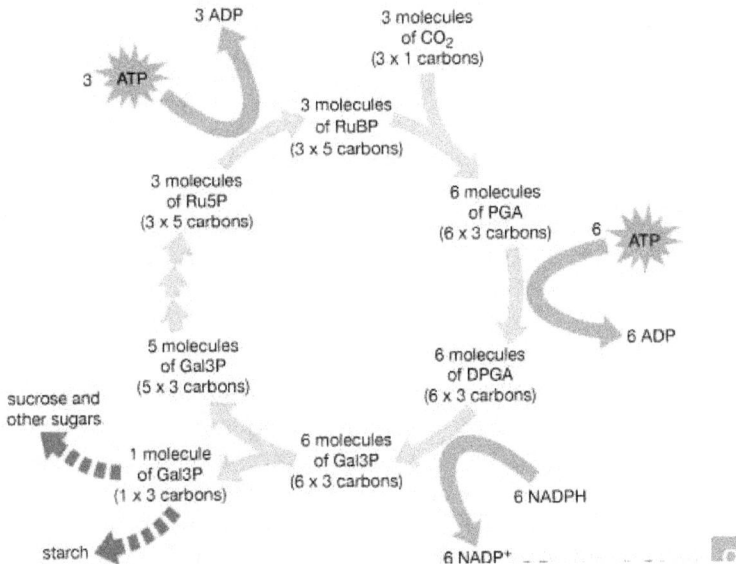

Pathway of carbon dioxide fixation and reduction in photosynthesis, the Calvin cycle. The diagram represents one complete turn of the cycle, with the net production of one molecule of glyceraldehyde-3-phosphate (Gal3P). This three-carbon sugar phosphate usually is converted to either sucrose or starch.

Evolutionary origins
The mitochondrion and chloroplast as independent entities

In addition to their remarkable metabolic capabilities, both mitochondria and chloroplasts synthesize on their own a number of proteins and lipids necessary for their structure and activity. Not only do they contain the machinery necessary for this, but they also possess the genetic material to direct it. DNA within these organelles has a circular structure reminiscent of prokaryotic, not eukaryotic, DNA. Also as in prokaryotes, the DNA is not associated with histones. Along with the DNA are protein-synthesizing ribosomes, of prokaryotic rather than eukaryotic size.

Only a small portion of the mitochondrion's total number of proteins is synthesized within the organelle. Numerous proteins are encoded and made in the cytoplasm specifically for export into the mitochondrion. The mitochondrial DNA itself encodes only 13 different proteins. The proteins that contain subunits synthesized within the mitochondrion often also possess subunits synthesized in the cytoplasm. Mitochondrial and chloroplastic proteins synthesized in the cytoplasm have to enter the organelle by a complex process, crossing both the outer and the inner membranes. These proteins contain specific arrangements of amino acids known as leader sequences that are recognized by receptors on the outer membranes of the organelles. The proteins are then guided through membrane channels in an energy-requiring process.

The endosymbiont hypothesis

Mitochondria and chloroplasts are self-dividing; they contain their own DNA and protein-synthesizing machinery, similar to that of prokaryotes. Chloroplasts produce ATP and trap photons by mechanisms that are complex and yet similar to those of certain prokaryotes. These phenomena have led to the theory that the two organelles are direct descendants of prokaryotes that entered primitive nucleated cells. Among billions of such events, a few could have led to the development of stable, symbiotic associations between nucleated hosts and prokaryotic parasites. The hosts would provide the parasites with a stable osmotic environment and easy access to nutrients, and the parasites would repay the hosts by providing an oxidative ATP-producing system or a photosynthetic energy-producing reaction.

The Cytoskeleton

The cytoskeleton is the name given to the fibrous network formed by different types of long protein filaments present throughout the cytoplasm of eukaryotic cells (cells containing a nucleus). The filaments of the cytoskeleton create a scaffold, or framework, that organizes other cell constituents and maintains the shape of the cell. In addition, some filaments cause coherent movements, both of the cell itself and of its internal organelles. Prokaryotic (nonnucleated) cells, which are generally much smaller than eukaryotic cells, contain a unique related set of filaments but, with few exceptions, do not possess true cytoskeletons. Their shapes and the shapes of certain eukaryotes, primarily yeast and other fungi, are determined by the rigid cell wall on the outside of the cell.

Four major types of cytoskeletal filaments are commonly recognized: actin filaments, microtubules, intermediate filaments, and septins. Actin filaments and microtubules are dynamic structures that continuously assemble and disassemble in most cells. Intermediate filaments are stabler and seem to be involved mainly in reinforcing cell structures, especially the position of the nucleus and the junctions that connect cells. Septins are involved in cell division and have been implicated in other cell functions. A wide variety of accessory proteins works in concert with each type of filament, linking filaments to one another and to the cell membrane and helping to form the networks that endow the cytoskeleton with its unique functions. Many of these accessory proteins have been characterized, revealing a rich diversity in the structure and function of the cytoskeleton.

Extranuclear DNA

All of the genetic information in a cell was initially thought to be confined to the DNA in the chromosomes of the cell nucleus. It is now known that small circular chromosomes, called extranuclear, or cytoplasmic, DNA, are located in two types of organelles found in the cytoplasm of the cell. These organelles are the mitochondria in animal and plant cells and the chloroplasts in plant cells. Chloroplast DNA (cpDNA) contains genes that are involved with aspects of photosynthesis and with components of the special protein-synthesizing apparatus that is active within the organelle. Mitochondrial DNA (mtDNA) contains some of the genes that participate in the conversion of the energy of chemical bonds into the energy currency of the cell—a chemical called adenosine triphosphate (ATP)—as well as genes for mitochondrial protein synthesis.

The cells of several groups of organisms contain small extra DNA molecules called plasmids. Bacterial plasmids are circular DNA molecules; some carry genes for resistance to various agents in the environment that would be toxic to the bacteria (e.g., antibiotics). Many fungi and some plants possess plasmids in their mitochondria; most of these are linear DNA molecules carrying genes that seem to be relevant only to the propagation of the plasmid and not the host cell.

Heredity And Evolution

At the centre of the theory of evolution as proposed by Charles Darwin and Alfred Russell Wallace were the concepts of variation and natural selection. Hereditary variants were thought to arise naturally in populations, and then these were either selected for or against by the contemporary environmental conditions. In this way, subsequent generations either became enriched or impoverished for specific variant types. Over the long term, the accumulation of such changes in populations could lead to the formation of new species and higher taxonomic categories. However, although hereditary change was basic to the theory, in the 19th-century world of Darwin and Wallace, the fundamental unit of heredity—the gene—was unknown. The birth and proliferation of the science of genetics in the 20th century after the discovery of Mendel's laws made it possible to consider the process of evolution by natural selection in terms of known genetic processes.

Population genetics

Because the processes of variation and selection take place at the population level, the basic theory of the genetics of evolutionary change is contained in the general area known as population genetics.

A simple way of viewing evolutionary change at the genotypic level would be to invent some hypothetical ancestral genotype, such as AAbbccDDEE, and an "evolved" derivative, such as aaBBccddee. (For illustrative purposes, only five genes are used, and these are assumed to be all homozygous.) Also, for simplicity it can be assumed that in both the ancestral and the evolved populations all individuals are identical. Clearly for all the genes except cc, a new allele completely replaces the original allele, and the new alleles can be either dominant or recessive. For example, in the case of the first gene, in the ancestral population all alleles are A, and in the evolved population all are a. For a to replace A, the population must go through stages in which there are mixtures of A and a alleles present in the population at the same time. In population genetics, allele frequency is the measurement of the commonness of an allele. The convention is to let the frequency of a dominant allele be p and that of a recessive allele q. Both are generally expressed as decimal fractions. In the above example, p changes from 1 to 0, and q changes from 0 to 1. Since there are only two alleles in this example, p + q must always equal 1. In the intermediate stages, there must be times when there are intermediate allele frequencies, for example when $p = 0.4$ and $q = 0.6$.

What can be said about genotype frequencies in the intermediate populations? In the ancestral and derived populations there must have been the following genotypic frequencies:

Ancestral AA = 1, Aa = 0, aa = 0
Evolved AA = 0, Aa = 0, aa = 1

Intuitively it seems that, in the intermediate stages, there must be more-complex proportions, including some heterozygotes. One possible intermediate stage is as follows:

AA = 0.30, Aa = 0.20, aa = 0.50

The allele frequencies at such an intermediate stage can be calculated by "adding up" the alleles. Hence, the frequency of A will be 0.30 plus 1/2 of 0.20 because the heterozygotes only carry one A allele. This is written

$p = 0.30 + 0.20/2 = 0.40$
Similarly, $q = 0.50 + 0.20/2 = 0.60$

(Noting these values for p and q, it is possible that this could have been the population discussed earlier, in which these specific values for p and q were hypothesized.)

In general, if D = frequency of homozygous dominants, R = frequency of homozygous recessives, and H = frequency of heterozygotes, then

$$p = D + H/2$$
and
$$q = R + H/2$$

This section has shown the importance of the concepts of allele frequency and genotype frequency in describing the genetic structure of populations. Of these, allele frequency is the simpler descriptor, and it forms the central tool of population genetics. Hence, the genetic basis of evolutionary change at the population level is described in terms of changes of allele frequencies.

Hardy-Weinberg equilibrium

It is a curious fact that populations show no inherent tendency to change allele or genotype frequencies. In the absence of selection or any of the other forces that can drive evolution, a population with given values of p and q will settle into a special stable set of genotypic proportions called a Hardy-Weinberg equilibrium. This principle was first realized by Godfrey Harold Hardy and Wilhelm Weinberg in 1908. The Hardy-Weinberg equilibrium of a population with allele frequencies p and q is defined by the set of genotypic frequencies p2 of AA, 2pq of Aa, and q2 of aa.

Monohybrid cross

Mother is heterozygous for a particular trait (Aa).

Father is also heterozygous for the same trait (Aa).

Homozygous dominant (AA) = 1/4
Heterozygous (Aa) = 1/2
Homozygous recessive (aa) = 1/4

♀\♂	A	a
A	AA	Aa
a	Aa	aa

Dihybrid cross (gene linkage)

A and a represent one trait, and B and b represent a different trait that is linked to inheritance of A or a.

	AB	Ab	aB	ab
AB	AABB	AABb	AaBB	AaBb
Ab	AABb	AAbb	AaBb	Aabb
aB	AaBB	AaBb	aaBB	aaBb
ab	AaBb	Aabb	aaBb	aabb

Dominant for A and B = 9/16
Recessive for a, dominant for B = 3/16
Dominant for A, recessive for b = 3/16
Recessive for a, recessive for b = 1/16

Punnett square diagrams are used to predict all the possible gene combinations that could result from the mating of parents with known genotypes for a particular trait. A monohybrid cross represents the inheritance pattern of a single trait, whereas a dihybrid cross represents the inheritance patterns of two traits that are linked.

When such a population reproduces itself to make a new generation, the lack of change is made apparent. It is intuitive that the allele frequencies p and q in the population are also measures of the frequencies of eggs and sperm used in creating a new generation (represented in the formula below). The new generation produced from the zygotes has exactly the same genotypic proportions as the first generation (the parents of the zygote).

Some specific allele frequencies, 0.7 for p and 0.3 for q, can be used to illustrate the calculation of the genotypic frequencies that constitute the Hardy-Weinberg equilibrium:

$p \times p = 0.7 \times 0.7 = 0.49$ of AA
$2 \times p \times q = 2 \times 0.7 \times 0.3 = 0.42$ of Aa
$q \times q = 0.3 \times 0.3 = 0.09$ of aa

When this population reproduces, there will be 0.49 + 0.21 = 0.7 of A gametes and 0.09 + 0.21 = 0.3 of a gametes, and, when these gametes combine, the population in the next generation will clearly have the same genotypic proportions as the previous one.

These simple calculations rely on several underlying assumptions. Perhaps the most crucial one is that there is random mating, or mating regardless of the genotype of the partner. In addition, the population must be large, and there can be no other pressures, such as selection, that can change allele frequencies. Despite these stringent requirements, many natural populations that have been studied are in Hardy-Weinberg equilibrium for the genes under investigation. The Hardy-Weinberg equilibrium constitutes an important benchmark for population genetic analysis.

If the Hardy-Weinberg principle of population genetics shows that there is no inherent tendency for evolutionary change, then how does change occur? This is considered in the following sections.

Changes in gene frequencies
Selection

One assumption behind the calculation of unchanging genotypic frequencies in Hardy-Weinberg equilibrium is that all genotypes have the same fitness. In genetics, fitness does not necessarily have to do with muscles; fitness is a measure of the ability to produce fertile offspring. In reality, the fitnesses of different genotypes are highly variable. The genotype with the greatest fitness is given the fitness value (w) of 1, and the lesser fitnesses are fractions of 1. For example, if snails of genotypes AA and Aa were to have an average of 100 offspring but those of genotype aa only 70, then the fitnesses of these three genotypes would be 1, 1, and 0.7, respectively. The proportional difference from the most fit is called the selection coefficient, s. Hence, s = 1 − w.

Alleles carried by less-fit individuals will be gradually lost from the population, and the relevant allele frequency will decline. This is the fundamental way in which natural selection operates in a population. Selection against dominant alleles is relatively efficient, because these are by definition expressed in the phenotype. Selection against recessive alleles is less efficient, because these alleles are sheltered in heterozygotes. Even though populations under selection technically are not in

Hardy-Weinberg equilibrium, the proportions of the formula can be used as an approximation to show the relative proportions of homozygous recessives and heterozygotes. If a rare deleterious recessive allele is of frequency 1/50 in the population, then (1/50)2, or 1 out of 2,500, individuals will express the recessive phenotype and be a candidate for negative selection. Heterozygotes will be at a frequency of 2pq = 2 × 49/50 × 1/50, or about 1 in 25. In other words, the heterozygotes are 100 times more common than recessive homozygotes; hence, most of the recessive alleles in a population will escape selection.

Because of the sheltering effect of heterozygotes, selection against recessive phenotypes changes the frequency of the recessive allele slowly. Even if the most severe level of selection is imposed, giving the recessive phenotype a fitness of zero (no fertile offspring), the recessive allele frequency (expressed as a fraction of the form 1/x) will increase in denominator by 1 in every generation. Therefore, to halve an allele frequency from 1/50 to 1/100 would proceed slowly from 1/50 to 1/51, 1/52, 1/53, and so on and would take 50 generations to get to 1/100. For lower intensities of selection, the progress would be even slower.

A different type of natural selection occurs when the fitness of a heterozygote exceeds the fitness of both homozygotes. The maintenance in human populations of the severe hereditary disease sickle cell anemia is owing to this form of selection. The disease allele (HbS) produces a specific type of hemoglobin that causes distortion (sickling) of the red blood cells in which the hemoglobin is carried. (Normal hemoglobin is coded by another allele, HbA). Accordingly, the possible genotypes are HbAHbA, HbAHbS, and HbSHbS. The latter individuals are homozygous for the sickle cell allele and will develop severe anemia because the oxygen transporting property of their blood is compromised. While the condition is not lethal before birth, such individuals rarely survive long enough to reproduce. On these grounds it might be expected that the disease allele would be selected against, driving the allele frequency to very low levels. However, in tropical areas of the world, the allele and the disease are common. The explanation is that the HbAHbS heterozygote is fitter and capable of leaving more offspring than is the homozygous normal HbAHbA in an environment containing the falciparum form of malaria. This extra measure of protection is evidently provided by the sickle cell hemoglobin, which is detrimental to the malaria parasite. In malarial environments, therefore, populations that contain the sickle cell gene have advantages over populations free of this gene. The former populations are in less danger from malaria, although they "pay" for this advantage by sacrificing in every generation some individuals who die of anemia.

Mutation

Genetics has shown that mutation is the ultimate source of all hereditary variation. At the level of a single gene whose normal functional allele is A, it is known that mutation can change it to a nonfunctional recessive form, a. Such "forward mutation" is more frequent than "back mutation" (reversion), which converts a into A. Molecular analysis of specific examples of mutant recessive alleles has shown that they are generally a heterogeneous set of small structural changes in the DNA, located throughout the segment of DNA that constitutes that gene. Hence, in an example from medical genetics, the disease phenylketonuria is inherited as a recessive phenotype and is ascribed to a causative allele that generally can be called k. However, sequencing alleles of many independent cases of phenylketonuria has shown that this k allele is in fact a set of many different kinds of mutational changes, which can be in any of the protein-coding regions of that gene.

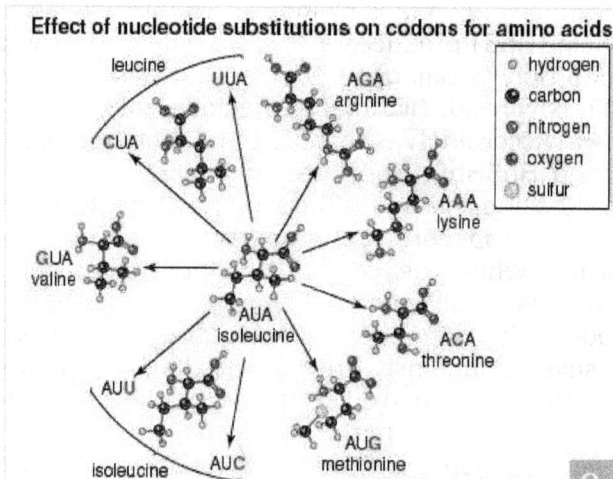

Effect of nucleotide substitutions on codons for amino acids

The effect of base substitutions, or point mutations, on the messenger-RNA codon AUA, which codes for the amino acid isoleucine. Substitutions (red letters) at the first, second, or third position in the codon can result in nine new codons corresponding to six different amino acids in addition to isoleucine itself. The chemical properties of some of these amino acids are quite different from those of isoleucine. Replacement of one amino acid in a protein by another can seriously affect the protein's biological function.

Recessive deleterious mutations are relatively rare, generally in the order of 1 per 105 or 106 mutant gametes per generation. Their constant occurrence over the generations, combined with the even greater rarity of back mutations, leads to a gradual accumulation in the population. This accumulation process is called mutational pressure.

Since mutational pressure to a deleterious recessive allele and selection pressure against the homozygous recessives are forces that act in opposite directions, another type of equilibrium is attained that effectively sets the value of q. Mathematically, q is determined by the following expression in which u is the net mutation rate of A to a, and s is the selection coefficient presented above:

$$q2 = (u/s), \text{ or } q = \text{Square root of} \sqrt{(u/s)}$$

Nonrandom mating

Many species engage in alternatives to random mating as normal parts of their cycle of sexual reproduction. An important exception is sexual selection, in which an individual chooses a mate on the basis of some aspect of the mate's phenotype. The selection can be based on some display feature such as bright feathers, or it may be a simple preference for a phenotype identical to the individual's own (positive assortative mating).

Two other important exceptions are inbreeding (mating with relatives) and enforced outbreeding. Both can shift the equilibrium proportions expected under Hardy-Weinberg calculations. For example, inbreeding increases the proportions of homozygotes, and the most extreme form of inbreeding, self-fertilization, eventually eliminates all heterozygotes.

Inbreeding and outbreeding are evolutionary strategies adopted by plants and animals living under certain conditions. Outbreeding brings gametes of different genotypes together, and the resulting individual differs from the parents. Increased levels of variation provide more evolutionary flexibility. All the showy colors and shapes of flowers are to promote this kind of exchange. In contrast, inbreeding maintains uniform genotypes, a strategy successful in stable ecological habitats.

In humans, various degrees of inbreeding have been practiced in different cultures. In most cultures today, matings of first cousins are the maximal form of inbreeding condoned by society. Apart from ethical considerations, a negative outcome of inbreeding is that it increases the likelihood of homozygosity of deleterious recessive alleles originating from common ancestors, called homozygosity by descent. The inbreeding coefficient F is a measure of the likelihood of homozygosity by descent; for example, in first-cousin marriages, $F = 1/16$. A large proportion of recessive hereditary diseases can be traced to first-cousin marriages and other types of inbreeding.

Random genetic drift

In populations of finite size, the genetic structure of a new generation is not necessarily that of the previous one. The explanation lies in a sampling effect, based on the fact that a subsample from any large set is not always representative of the larger set. The gametes that form any generation can be thought of as a sample of the alleles from the parental one. By chance the sample might not be random; it could be skewed in either direction. For example, if $p = 0.600$ and $q = 0.400$, sampling "error" might result in the gametes having a p value of 0.601 and a q of 0.399. If by chance this skewed sampling occurs in the same direction from generation to generation, the allele frequency can change radically. This process is known as random genetic drift. As might be expected, the smaller the population, the greater chance of sampling error and hence significant levels of drift in any one generation. In extreme cases, drift over the generations can result in the complete loss of one allele; in these occurrences the other is said to be fixed.

Other cases of sampling error occur when new colonies of plants or animals are founded by small numbers of migrants (founder effect) and when there is radical reduction in population size because of a natural catastrophe (population bottleneck). One inevitable effect of these processes is a reduction in the amount of variation in the population after the size reduction. Two species that have gone through drastic bottlenecks with the associated reduction of genetic variation are cheetahs (Africa) and northern elephant seals (North America).

Microevolution

There is ample evidence that the processes described above are at work in natural populations. Together, these changes are called microevolution—in other words, small-scale evolution. Even within the relatively short period of time since Darwin, it has been possible to document such processes. Allelic variation has been found to be common in nature. It is detected as polymorphism, the presence of two or more distinct hereditary forms associated with a gene. Polymorphism can be morphological, such as blue and brown forms of a species of marine mussel, or molecular, detectable only at the DNA or protein level. Although much of this polymorphism is not understood, there are enough examples of selection of polymorphic forms to indicate that it is potentially adaptive. Selection has been observed favouring melanic (dark) forms of peppered moths in industrial areas and favouring resistance to toxic agents such as the insecticide DDT, the rat poison warfarin, and the virus that causes the disease myxomatosis in rabbits.

Industrial Melanism In The Peppered Moth

The light gray form of the peppered moth (Biston betularia) is inconspicuous on the lichen-covered tree trunk. In contrast, the dark (melanic) morph stands out, leaving it vulnerable to predation by birds. The gradual darkening of the wings of the melanic peppered moth, the result of living in woodlands darkened by industrial pollution, is an example of industrial melanism.

From the experiments of Dr. H.B.D. Kettlewell, University of Oxford; photographs by John S. Haywood

More-complex genetic changes have been documented, leading to special locally adapted "ecotypes." Anoles (a type of lizard) on certain Caribbean islands show convincing examples of adaptations to specific habitats, such as tree trunks, tree branches, or grass. Introductions of lizards onto uncolonized islands result in demonstrable microevolutionary adaptations to the various vacant niches. On the Galapagos Islands, studies over several decades have documented adaptive changes in the beaks of finches. In some studies, documented changes have led to incipient new species. An example is the apple maggot, the larva of a fly in North America that has evolved from a similar fly living on hawthorns—all in the period since the introduction of apples. The formation of new species was a key component of Darwin's original theory. Now it appears that the accumulation of enough small-scale genetic changes can lead to the inability to mate with members of an ancestral population; such reproductive isolation is the key step in species formation.

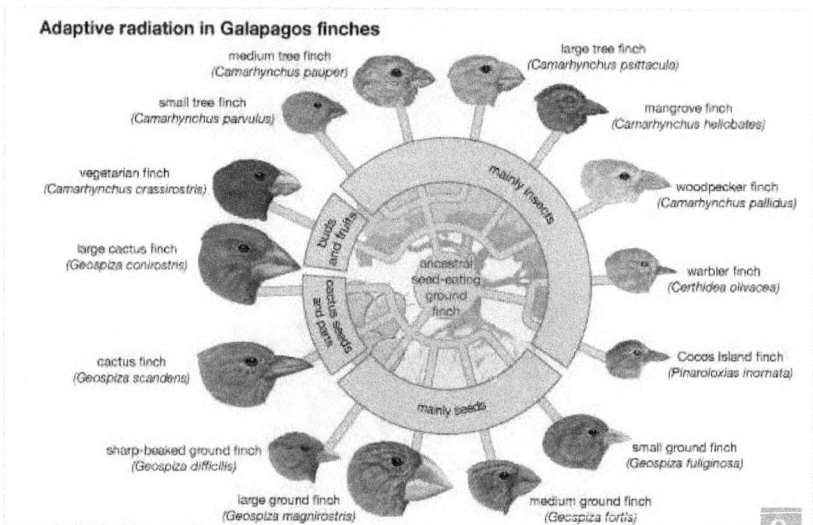

Adaptive radiation in Galapagos finches

medium tree finch
(Camarhynchus pauper)

large tree finch
(Camarhynchus psittacula)

small tree finch
(Camarhynchus parvulus)

mangrove finch
(Camarhynchus heliobates)

vegetarian finch
(Camarhynchus crassirostris)

woodpecker finch
(Camarhynchus pallidus)

large cactus finch
(Geospiza conirostris)

warbler finch
(Certhidea olivacea)

cactus finch
(Geospiza scandens)

Cocos Island finch
(Pinaroloxias inornata)

sharp-beaked ground finch
(Geospiza difficilis)

small ground finch
(Geospiza fuliginosa)

large ground finch
(Geospiza magnirostris)

medium ground finch
(Geospiza fortis)

mainly insects · buds and fruit · cactus and pear · ancestral seed-eating ground finch · mainly seeds

Fourteen species of Galapagos finches that evolved from a common ancestor. The different shapes of their bills, suited to different diets and habitats, show the process of adaptive radiation.

It is reasonable to assume that the continuation of microevolutionary genetic changes over very long periods of time can give rise to new major taxonomic groups, the process of macroevolution. There are few data that bear directly on the processes of macroevolution, but gene analysis does provide a way for charting macroevolutionary relationships indirectly.

DNA phylogeny

The ability to isolate and sequence specific genes and genomes has been of great significance in deducing trees of evolutionary relatedness. An important discovery that enables this sort of analysis is the considerable evolutionary conservation between organisms at the genetic level. This means that different organisms have a large proportion of their genes in common, particularly those that code for proteins at the central core of the chemical machinery of the cell. For example, most organisms have a gene coding for the energy-producing protein cytochrome C, and furthermore, this gene has a very similar nucleotide sequence in all organisms (that is, the sequence is conserved). However, the sequences of cytochrome C in different organisms do show differences, and the key to phylogeny is that the differences are proportionately fewer between organisms that are closely related. The interpretation of this observation is that organisms that share a common ancestor also share common DNA sequences derived from that ancestor. When one ancestral species splits into two, differences accumulate as a result of mutations, a process called divergence. The greater the amount of divergence, the longer must have been the time since the split occurred. To carry out this sort of analysis, the DNA sequence data are fed into a computer. The computer positions similar species together on short adjacent branches showing a relatively recent split and dissimilar species on long branches from an ancient split. In this way a molecular phylogenetic tree of any number of organisms can be drawn.

Phylogeny based on nucleotide differences in the gene for cytochrome c

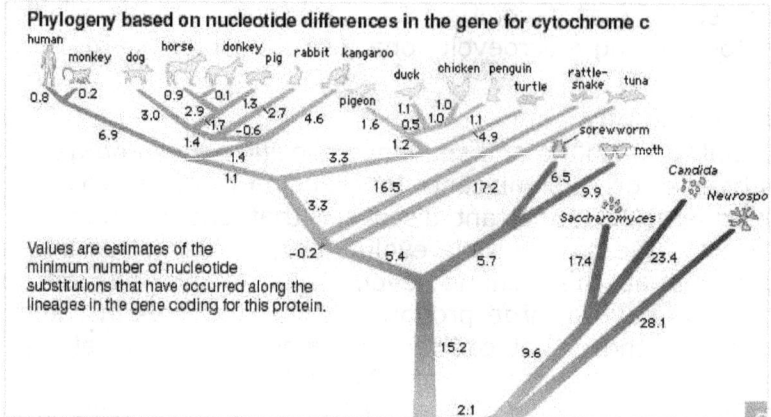

Figure : Phylogeny based on differences in the protein sequence of cytochrome c in organisms ranging from Neurospora mold to humans.

DNA difference in some cases can be correlated with absolute dates of divergence as deduced from the fossil record. Then it is possible to calculate divergence as a rate. It has been found that divergence is relatively constant in rate, giving rise to the idea that there is a type of "molecular clock" ticking in the course of evolution. Some ticks of this clock (in the form of mutations) are significant in terms of adaptive changes to the gene, but many are undoubtedly neutral, with no significant effect on fitness.

One of the interesting discoveries to emerge from molecular phylogeny is that gene duplication has been common during evolution. If an extra copy of a gene can be made, initially by some cellular accident, then the "spare" copy is free to mutate and evolve into a separate function.

Molecular phylogeny of some genes has also pointed to unexpected cases of, say, a plant gene nested within a tree of animal genes of that type or a bacterial gene nested within a plant phylogenetic tree. The explanation for such anomalies is that there has been horizontal transmission from one group to another. In other words, on rare occasions a gene can hop laterally from one species to another. Although the mechanisms for horizontal transmission are presently not known, one possibility is that bacteria or viruses act as natural vectors for transferring genes.

Synteny

Genomic sequencing and mapping have enabled comparison of the general structures of genomes of many different species. The general finding is that organisms of relatively recent divergence show similar blocks of genes in the same relative positions in the genome. This situation is called synteny, translated roughly as possessing common chromosome sequences. For example, many of the genes of humans are syntenic with those of other mammals—not only apes but also cows, mice, and so on. Study of synteny can show how the genome is cut and pasted in the course of evolution.

Polyploidy

Genomic analysis also has shown that one of the important mechanisms of evolution is multiplication of chromosome sets, resulting in polyploidy ("many genomes"). In plants and animals, spontaneous doubling of chromosomes can occur. In some plants, the chromosomes of two related species unite via cross-pollination to form a fusion product. This product is sterile because each chromosome needs a pairing partner in order for the plant to be fertile. However, the chromosomes of the fusion product can accidentally double, resulting in a new, fertile species. Wheat is an example of a plant that evolved by this means through a union between wild grasses, but a large proportion of plants went through similar ancestral polyploidization.

Human evolution

Many of the techniques of evolutionary genetics can be applied to the evolution of humans. Charles Darwin created a large controversy in Victorian England by suggesting in his book The Descent of Man that humans and apes share a common ancestor. Darwin's assertion was based on the many shared anatomical features of apes and humans. DNA analysis has supported this hypothesis. At the DNA sequence level, the genomes of humans and chimpanzees are 99 percent identical. Furthermore, when phylogenetic trees are constructed using individual genes, humans and apes cluster together in short terminal branches of the trees, suggesting very recent divergence. Synteny too is impressive, with relatively minor chromosomal rearrangements.

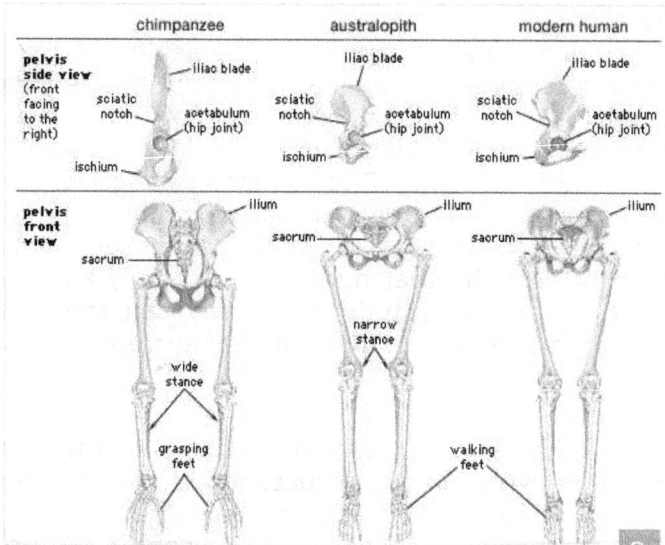

Comparison of the pelvis and lower limbs
of a chimpanzee, an australopith, and a modern human.

Fossils have been found of various extinct forms considered to be intermediates between apes and humans. Notable is the African genus Australopithecus, generally believed to be one of the earliest hominins and an intermediate on the path of human evolution. The first toolmaker was Homo habilis, followed by Homo erectus and finally Homo sapiens (modern humans). H. habilis fossils have been found only in Africa, whereas fossils of H. erectus and H. sapiens are found throughout the Old World. Phylogenetic trees based on DNA sequencing of all peoples have shown that Africans represent the root of the trees. This is interpreted as evidence that H. sapiens evolved in Africa, spread throughout the globe, and outcompeted H. erectus wherever the two cohabited.

Variations of DNA, either unique alleles of individual genes or larger-sized blocks of variable structure, have been used as markers to trace human migrations across the globe. Hence, it has been possible to trace the movement of H. sapiens out of Africa and into Europe and Asia and, more recently, to the American continents. Also, genetic markers are useful in plotting human migrations that occurred in historical time. For example, the invasion of Europe by various Asian conquerors can be followed using blood-type alleles.

As humans colonized and settled permanently in various parts of the world, they differentiated themselves into distinct groups called races. Undoubtedly, many of the features that distinguish races, such as skin colour or body shape, were adaptive in the local settings, although such adaptiveness is difficult to demonstrate. Nevertheless, genomic analysis has revealed that the concept of race has little meaning at the genetic level. The differences between races are superficial, based on the alleles of a relatively small number of genes that affect external features. Furthermore, while races differ in allele frequencies, these same alleles are found in most races. In other words, at the genetic level there are no significant discontinuities between races. It is paradoxical that race, which has been so important to people throughout the course of human history, is trivial at the genetic level—an important insight to emerge from genetic analysis.

Actin Filaments

Actin is a globular protein that polymerizes (joins together many small molecules) to form long filaments. Because each actin subunit faces in the same direction, the actin filament is polar, with different ends, termed "barbed" and "pointed." An abundant protein in nearly all eukaryotic cells, actin has been extensively studied in muscle cells. In muscle cells, the actin filaments are organized into regular arrays that are complementary with a set of thicker filaments formed from a second protein called myosin. These two proteins create the force responsible for muscle contraction. When the signal to contract is sent along a nerve to the muscle, the actin and myosin are activated. Myosin works as a motor, hydrolyzing adenosine triphosphate (ATP) to release energy in such a way that a myosin filament moves along an actin filament, causing the two filaments to slide past each other. The thin actin filaments and the thick myosin filaments are organized in a structure called the sarcomere, which shortens as the filaments slide over one another. Skeletal muscles are composed of bundles of many long muscle cells; when the sarcomeres contract, each of these giant muscle cells shortens, and the overall effect is the contraction of the entire muscle. Although the stimulation pathways differ, heart muscle and smooth muscle (found in many internal organs and blood vessels) contract by a similar sliding filament mechanism.

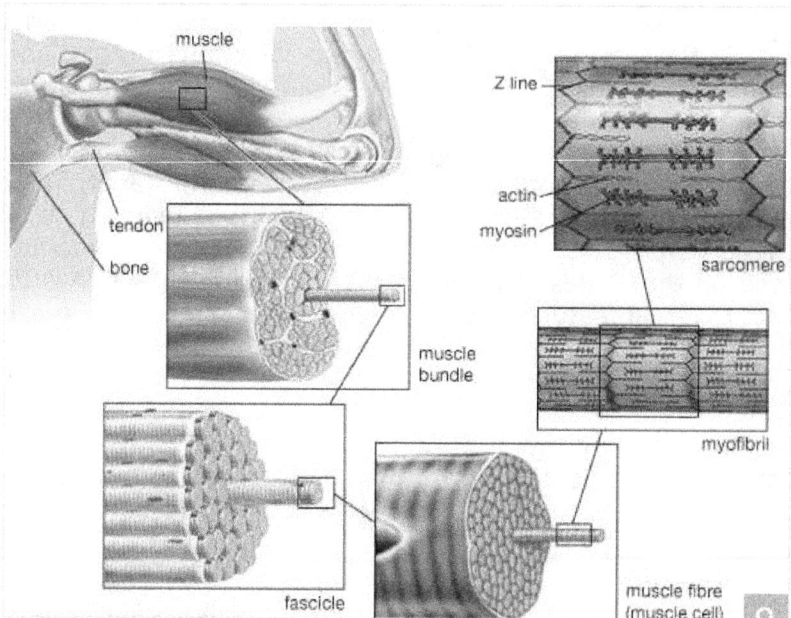

The structure of striated muscleStriated muscle tissue, such as the tissue of the human biceps muscle, consists of long, fine fibres, each of which is in effect a bundle of finer myofibrils. Within each myofibril are filaments of the proteins myosin and actin; these filaments slide past one another as the muscle contracts and expands. On each myofibril, regularly occurring dark bands, called Z lines, can be seen where actin and myosin filaments overlap. The region between two Z lines is called a sarcomere; sarcomeres can be considered the primary structural and functional unit of muscle tissue.

Actin is also present in non-muscle cells, where it forms a meshwork of filaments responsible for many types of cellular movement. The meshwork consists of actin filaments that are attached to the cell membrane and to each other. The length of the filaments and the architecture of their attachments determine the shape and consistency of a cell. A large number of accessory proteins bind to actin, controlling the number, length, position, and attachments of the actin filaments. Different cells and tissues contain different accessory proteins, which accounts for the different shapes and movements of different cells. For example, in some cells, actin filaments are bundled by accessory proteins, and the bundle is attached to the cell membrane to form microvilli, stable protrusions that resemble tiny bristles. Microvilli on the surface of epithelial cells such as those lining the intestine increase the cell's surface area and thus facilitate the absorption of ingested food and water molecules. Other types of microvilli are involved in the detection of sound in the ear, where their movement, caused by sound waves, sends an electrical signal to the brain.

Many actin filaments in non-muscle cells have only a transient existence, polymerizing and depolymerizing in controlled ways that create movement. For example, many cells continually send out and retract tiny "filopodia," long needlelike projections of the cell membrane that are thought to enable cells to probe their environment and decide which direction to go. Like microvilli, filopodia are formed when actin filaments push out the membrane, but, because these actin filaments are less stable, filopodia have only a brief existence. Another actin structure only transiently associated with the cell membrane is the contractile ring, which is composed of actin filaments running around the circumference of the cell during cell division. As its name implies, this ring pulls in the cell membrane by a myosin-dependent process, thereby pinching the cell in half.

Microtubules

Microtubules are long filaments formed from 13 to 15 protofilament strands of a globular subunit called tubulin, with the strands arranged in the form of a hollow cylinder. Like actin filaments, microtubules are polar, having "plus" and "minus" ends. Most microtubule plus ends are constantly growing and shrinking, by respectively adding and losing subunits at their ends. Stable microtubules are found in cilia and flagella. Cilia are hairlike structures found on the surface of certain types of epithelial cells, where they beat in unison to move fluid and particles over the cell surface. Cilia are closely related in structure to flagella. Flagella such as those found on sperm cells produce a helical wavelike motion that enables a cell to propel itself rapidly through fluids. In cilia and flagella a set of microtubules is connected in a regular array by numerous accessory proteins that act as links and spokes in the assembly. Movement of the cilia or flagella occurs when adjacent microtubules slide past one another, bending the structures. This motion is caused by the motor protein dynein, which uses the energy of ATP hydrolysis to move along the microtubules, in a manner resembling the movement of myosin along actin filaments.

In most cells, microtubules grow outward, from the cell centre to the cell membrane, from a special region of the cytoplasm near the nuclear envelope called the centrosome. The minus ends of these microtubules are embedded in the centrosome, while the plus ends terminate near the cell membrane. The plus ends grow and shrink rapidly, a process known as dynamic instability. At the start of cell division, the centrosome replicates and divides in two. The two centrosomes separate and move to opposite sides of the nuclear envelope, where each nucleates a starlike array of microtubules, forming the mitotic spindle. The mitotic spindle partitions the duplicated chromosomes into the two daughter cells during mitosis.

Microtubules often serve as tracks for the transport of membrane vesicles in the cell, carried by the motor proteins kinesin and dynein. Kinesins generally move toward the plus end of the microtubule, and dyneins move toward the minus end. Microtubule-based vesicle transport occurs in nearly all cells, but it is especially prominent in the long thin processes of neurons, carrying essential components to and from the synapses at the ends of the processes.

Intermediate Filaments

Intermediate filaments are so named because they are thicker than actin filaments and thinner than microtubules or muscle myosin filaments. The subunits of intermediate filaments are elongated, not globular, and are associated in an antipolar manner. As a result, the overall filament has no polarity, and therefore no motor proteins move along intermediate filaments. Intermediate filaments are found only in complex multicellular organisms. They are encoded by a large number of different genes and can be grouped into families based on their amino acid sequences. Cells in different tissues of the body express one or another of these genes at different times. One cell can even change which type of intermediate filament protein is expressed over its lifetime. Most likely, the different forms of intermediate filaments have subtle but critical differences in their functional characteristics, helping to define the function of the cell. In general, intermediate filaments serve as structural elements, helping cells maintain their shape and integrity. For example, keratin filaments, the intermediate filaments of epithelial cells, which line surfaces of the body, give strength to the cell sheet that covers the surface. Mutations in keratin genes can result in blisters when the epithelial cell sheet is weak and prone to rupture. Keratin mutations can also cause deformations in the hair, nails, and corneas. Another example of a family of intermediate filaments is the lamin family, which comprises the nuclear lamina, a fibrous shell that underlies and supports the nuclear membrane.

The Cell Matrix And Cell-To-Cell Communication
The development of single cells into multicellular organisms involves a number of adaptations. The cells become specialized, acquiring distinct functions that contribute to the survival of the organism. The behaviour of individual cells is also integrated with that of similar cells, so that they act together in a regulated fashion. To achieve this integration, cells assemble into specialized tissues, each tissue being composed of cells and the spaces outside of the cells.

The surface of cells is important in coordinating their activities within tissues. Embedded in the plasma membrane of each cell are a number of proteins that interact with the surface or secretions of other cells. These proteins enable cells to "recognize" and adhere to the extracellular matrix and one another and to form populations distinct from surrounding cells. These interactions are key to the organizational behaviour of cell populations and contribute to the formation of embryonic tissues and the function of normal tissue in the adult organism.

The Extracellular Matrix
A substantial part of tissues is the space outside of the cells, called the extracellular space. This is filled with a composite material, known as the extracellular matrix, composed of a gel in which a number of fibrous proteins are suspended. The gel consists of large polysaccharide (complex sugar) molecules in a water solution of inorganic salts, nutrients, and waste products known as the interstitial fluid. The major types of protein in the matrix are structural proteins and adhesive proteins.

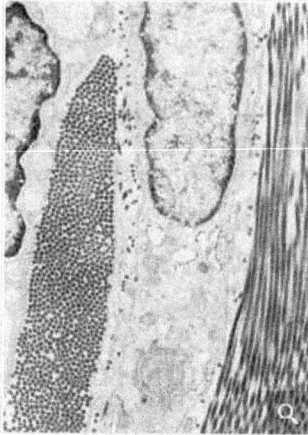

Electron micrograph of a small area of dense fibrous connective tissue, illustrating the intimate association of cells and fibres. In the centre is a portion of a fibrocyte, and on either side are two collagen fibres. The collagen fibre on the left is cut transversely, showing round cross sections of the unit fibrils. The collagen fibre on the right has been cut nearly parallel to its long axis and shows extensive segments of the cross-striated fibrils.

There are two general types of tissues distinct not only in their cellular organization but also in the composition of their extracellular matrix. The first type, mesenchymal tissue, is made up of clusters of cells grouped together but not closely adherent to one another. They synthesize a highly hydrated gel, rich in salts, fluid, and fibres, known as the interstitial matrix. Connective tissue is a mesenchyme that fastens together other more highly organized tissues. The solidity of various connective tissues varies according to the consistency of their extracellular matrix, which in turn depends on the water content of the gels, the amount and type of polysaccharides and structural proteins, and the presence of other salts. For example, bone is rich in calcium phosphate, giving that tissue its rigidity; tendons are mostly fibrous structural proteins, yielding a ropelike consistency; and joint spaces are filled with a lubricating fluid of mostly polysaccharide and interstitial fluid.

Epithelial tissues, the second type, are sheets of cells adhering at their side, or lateral, surfaces. They synthesize and deposit at their bottom, or basal, surfaces an organized complex of matrix materials known as the basal lamina or basement membrane. This thin layer serves as a boundary with connective tissue and as a substrate to which epithelial cells are attached.

Matrix Polysaccharides
The polysaccharides, or glycans, of the extracellular matrix are responsible for its gel-like quality and for organizing its components. These large acidic molecules exist alone (as glycosaminoglycans) or in combination with small proteins (as proteoglycans). They bind an extraordinarily large amount of water, thus forming massively swollen gels that fill the spaces between cells. Bound to proteins, they also organize other molecules in the extracellular matrix. The firmness and resiliency of cartilage, as at the surface of joints, is due to highly organized proteoglycans that bind water tightly.

Matrix Proteins
Matrix proteins are large molecules tightly bound to form extensive networks of insoluble fibres. These fibres may even exceed the size of the cells themselves. The proteins are of two general types, structural and adhesive.

The structural proteins, collagen and elastin, are the dominant matrix proteins. At least 10 different types of collagen are present in various tissues. The most common, type I collagen, is the most abundant protein in vertebrate animals, accounting for nearly 25 percent of the total protein in the body. The various collagen types share structural features, all being composed of three intertwined polypeptide chains. In some collagens the chains are linked together by covalent bonds, yielding a ropelike structure of great tensile strength. Indeed, the toughness of leather, chemically treated animal skin, is due to its content of collagen. Elastin is also a cross-linked protein, but, instead of forming rigid coils, it imparts elasticity to tissues. Only one type of elastin is known; it varies in elasticity according to variations in its cross-linking.

The adhesive proteins of the extracellular matrix bind matrix molecules to one another and to cell surfaces. These proteins are modular in that they contain several functional domains packaged together in a single molecule. Each domain binds to a specific matrix component or to a specific site on a cell. The major adhesive protein of the interstitial matrix is called fibronectin; that of the basal lamina is known as laminin.

Cell-Matrix Interactions

Molecules intimately associated with the cell membrane link cells to the extracellular matrix. These molecules, called matrix receptors, bind selectively to specific matrix components and interact, directly or indirectly, with actin protein fibres that form the cytoskeleton inside the cell. This association of actin fibres with matrix components via receptors on the cell membrane can influence the organization of membrane molecules as well as matrix components and can modify the shape and function of the cytoskeleton. Changes in the cytoskeleton can lead to changes in cell shape, movement, metabolism, and development.

Intercellular Recognition And Cell Adhesion

The ability of cells to recognize and adhere to one another plays an important role in cell survival and reproduction. For example, when starved, several types of single-cell organisms band together to develop the specialized cells needed for reproduction. In this process, certain cells at the centre of the developing aggregate secrete chemicals that cause the other cells to adhere tightly into a group. In the case of slime mold amoebas, starvation causes the secretion of a compound, cyclic adenosine monophosphate (cyclic AMP, or CAMP), that induces the cells to stick together end to end. With further aggregation, the cells produce another cell-surface glycoprotein with which they stick to one another over their entire surfaces. The cellular aggregates then produce an extracellular matrix, which holds the cells together in a specific structural form.

Tissue And Species Recognition

Some multicellular animals or tissues can be dissociated into suspensions of single cells that show the same cellular recognition and adhesion as do aggregates of single-cell organisms. The marine sponge, for example, can be sieved through a mesh, yielding single cells and cells in clumps. When this cell suspension is rotated in culture, the cells reaggregate and in time reform a normal sponge. This reassociation shows selective cell recognition; that is, only cells of the same species reassociate. The ability of the cells to distinguish cells of their own species from those of others is mediated by proteoglycan molecules in the extracellular matrix. The proteoglycan binds to specific cell-surface receptor sites that are unique to a single species of sponge.

Marine sponges are multicellular animals that can regenerate from single cells. The cells of a sponge rely on the processes of intercellular recognition and cellular adhesion to form aggregates of cells of the same species that eventually develop into an adult sponge.

Cells from tissues of vertebrate animals can, like sponge cells, be dissociated and allowed to reaggregate. For example, when vertebrate embryonic cells from two different tissues are dissociated and then rotated together in culture, the cells form a multicellular aggregate within which they sort according to the type of tissue, a sorting that occurs regardless of whether the cells are from the same or different species. The specificity is due to a set of cell-surface glycoproteins called cell adhesion molecules (CAM). A portion of the CAM that extends from the surface of a cell adheres to identical molecules on the surface of adjacent cells. These CAM appear early in embryonic life, and their amounts in tissues change as the organs develop. The CAM, however, are not responsible for the stable adhesion of one cell to another; this more permanent adhesion is carried out by cell junctions.

Cell Junctions
There are three functional categories of cell junction: adhering junctions, often called desmosomes; tight, or occluding, junctions; and gap, or permeable, junctions. Adhering junctions hold cells together mechanically and are associated with intracellular fibres of the cytoskeleton. Tight junctions also hold cells together, but they form a nearly leakproof intercellular seal by fusion of adjacent cell membranes. Both adhering

junctions and tight junctions are present primarily in epithelial cells. Many cell types also possess gap junctions, which allow small molecules to pass from one cell to the next through a channel.

Adhering Junctions

Cells subject to abrasion or other mechanical stress, such as those of the surface epithelia of the skin, have junctions that adhere cells to one another and to the extracellular matrix. These adhering junctions are called desmosomes when occurring between cells and hemidesmosomes (half-desmosomes) when linked to the matrix. Adhering junctions distribute mechanical shear force throughout the tissue and to the underlying matrix by virtue of their association with intermediate filaments crossing the interior of the cell. The linkage of these filaments, also called keratin filaments, to the desmosomes and, through these junctions, to adjacent cells provides a nearly continuous fibrous network throughout an epithelial sheet. Adhering junctions are also seen in other types of cells—for example, in the muscles of the heart and uterus—allowing these cells to remain anchored together despite the contractions of the muscles.

Tight Junctions

Sheets of cells separate fluids within the organs from fluids outside, as in the epithelial layer lining the intestine. This separation requires leakproof junctions between cells. Tight junctions form leakproof seals by fusing the plasma membranes of adjacent cells, creating a continuous barrier through which molecules cannot pass. The membranes are fused by tight associations of two types of specialized integral membrane proteins, in turn repelling large water-soluble molecules. In invertebrates this function is provided by septate junctions, in which the proteins of the membrane rather than the lipids form the seal.

Gap Junctions

These junctions allow communication between adjacent cells via the passage of small molecules directly from the cytoplasm of one cell to that of another. Molecules that can pass between cells coupled by gap junctions include inorganic salts, sugars, amino acids, nucleotides, and vitamins but not large molecules such as proteins or nucleic acids.

Gap junctions are crucial to the integration of certain cellular activities. For example, heart muscle cells generate electrical current by the movement of inorganic salts. If the cells are coupled, they will share this electrical current, allowing the synchronous contraction of all the cells in the tissue. This coupling function requires the regulation of molecular traffic through the gaps. The junctions are not open pores but dynamic channels, which change their permeability with changes in cellular activity. They consist of proteins completely crossing the cell membrane as six-sided columns with central pores. Under certain conditions the proteins are thought to change shape, causing the pores to become smaller or larger and thus changing the permeability of the junction.

Gap junctions are also found in tissues that are not electrically active. In these tissues, the junctions allow nutrients and waste products to travel throughout the tissue. Cells in such tissues are said to be metabolically coupled. During the formation of embryos, gap junctions are crucial to establishing differences between separate groups of cells, the coupled cells undergoing development together to become a specialized tissue.

Cell-to-Cell Communication via Chemical Signaling

In addition to cell-matrix and cell-cell interactions, cell behaviour in multicellular organisms is coordinated by the passage of chemical or electrical signals between cells. The most common form of chemical signaling is via molecules secreted from the cells and moving through the extracellular space. Signaling molecules may also remain on cell surfaces, influencing other cells only after the cells make physical contact. Finally, as noted above, gap junctions allow small molecules to move between the cytoplasms of adjacent cells.

Types of Chemical Signaling

Chemical signals secreted by cells can act over varying distances. In the autocrine signaling process, molecules act on the same cells that produce them. In paracrine signaling, they act on nearby cells. Autocrine signals include extracellular matrix molecules and various factors that stimulate cell growth. An example of paracrine signals is the chemical transmitted from nerve to muscle that causes the muscle to contract. In this instance, the muscle cells have regions specialized to receive chemical signals from an adjacent nerve cell. In both autocrine and paracrine signaling, the chemical signal works in the immediate vicinity of the cell that produces it and is present at high concentrations. A chemical signal picked up by the bloodstream and taken to distant sites is called an endocrine signal. Most hormones produced in vertebrates are endocrine signals, such as the hormones produced in the pituitary gland at the base of the brain and carried by the bloodstream to act at low concentrations on the thyroid or adrenal glands.

The concentration at which a chemical signal acts has significance for its target cell. Chemical signals that act at high concentration act locally and rapidly. On the other hand, chemical signals that act at low concentrations act at distances and are generally slow.

Signal Receptors

The ability of a cell to respond to an extracellular signal depends on the presence of specific proteins called receptors, which are located on the cell surface or in the cytoplasm. Receptors bind chemical signals that ultimately trigger a mechanism to modify the behaviour of the target cell. Cells may contain an array of specific receptors that allow them to respond to a variety of chemical signals.

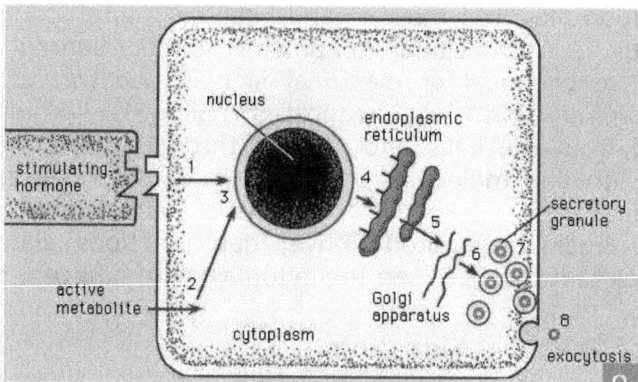

Hormones and active metabolites bind to different types of receptors. Water-soluble molecules (i.e., insulin) cannot pass through the lipid membrane of a cell and thus rely on cell surface receptors to transmit messages to the interior of the cell. In contrast, lipid-soluble molecules (i.e., certain active metabolites) are able to diffuse through the lipid membrane to communicate messages directly to the nucleus.

Hormones and active metabolites bind to different types of receptors. Water-soluble molecules (i.e., insulin) cannot pass through the lipid membrane of a cell and thus rely on cell surface receptors to transmit messages to the interior of the cell. In contrast, lipid-soluble molecules (i.e., certain active metabolites) are able to diffuse through the lipid membrane to communicate messages directly to the nucleus.

Signal molecules are either soluble or insoluble. Water-soluble molecules, such as the polypeptide hormone insulin, bind to receptors at cell surfaces. On the other hand, lipid-soluble molecules, such as the steroid hormones produced by the ovary or testis, pass through the lipid bilayer of the cell membrane to reach receptors within the cytoplasm. Extracellular matrix molecules are chemical signals, but, because of their size and insolubility, they act only on cell surface receptors and are neither taken up by the cells nor rapidly destroyed.

Cellular Response
The binding of chemical signals to their corresponding receptors induces events within the cell that ultimately change its behaviour. The nature of these intracellular events differs according to the type of receptor. Also, the same chemical signal can trigger different responses in different types of cell.

Cell surface receptors work in several ways when they are occupied. Some receptors enter the cell still bound to the chemical signal. Others activate membrane enzymes, which produce certain intracellular chemical mediators. Still other receptors open membrane channels, allowing a flow of ions that causes either a change in the electrical properties of the membrane or a change in the ion concentration in the cytoplasm. This regulation of enzymes or membrane channels produces changes in the concentration of intracellular signaling molecules, which are often called second messengers (the first messenger being the extracellular chemical signal bound to the receptor).

Two common intracellular signaling molecules are cyclic AMP and the calcium ion. Cyclic AMP is a derivative of adenosine triphosphate, the ubiquitous energy-carrying molecule of the cell. The intracellular concentrations of both cyclic AMP and calcium ions are normally very low. The binding of an extracellular chemical signal to a cell surface receptor stimulates an enzyme complex in the membrane to produce cyclic AMP. This second messenger then diffuses into the cytoplasm and acts on intracellular enzymes called kinases that modify the behaviour of the cell, culminating in the activation of target genes that increase the synthesis of certain proteins. The action of cyclic AMP is brief because it is rapidly degraded by specific enzymes.

Occupancy of other surface receptors causes a transient opening of membrane channels. This can allow calcium ions to enter the cytoplasm from the extracellular space, where their concentration is higher. The action of calcium ions is also brief because they are rapidly pumped out of the cell or bound to intracellular molecules, lowering the cytoplasmic concentration to the state existing before stimulation.

Some extracellular chemical signals enter the cell intact, still bound to the receptor, without generating a second messenger. In this mechanism, receptor occupancy causes individual receptors within the cell membrane to aggregate spontaneously. That portion of the membrane containing the aggregated receptors is then taken into the cell, where it fuses with various membrane-bounded organelles in the cytoplasm. In some instances the chemical signal is released within the organelles, and in almost all instances the ingested membrane is rapidly returned to the cell membrane along with the surface receptors.

The Plant Cell Wall
The plant cell wall is a specialized form of extracellular matrix that surrounds every cell of a plant and is responsible for many of the characteristics distinguishing plant from animal cells. Although often perceived as an inactive product serving mainly mechanical and structural purposes, the cell wall actually has a multitude of functions upon which plant life depends. Such functions include: (1) providing the protoplast, or living cell, with mechanical protection and a chemically buffered environment, (2) providing a porous medium for the circulation and distribution of water, minerals, and other small nutrient molecules, (3) providing rigid building blocks from which stable structures of higher order, such as leaves and stems, can be produced, and (4) providing a storage site of regulatory molecules that sense the presence of pathogenic microbes and control the development of tissues.

Mechanical Properties of Wall Layers

All cell walls contain two layers, the middle lamella and the primary cell wall, and many cells produce an additional layer, called the secondary wall. The middle lamella serves as a cementing layer between the primary walls of adjacent cells. The primary wall is the cellulose-containing layer laid down by cells that are dividing and growing. To allow for cell wall expansion during growth, primary walls are thinner and less rigid than those of cells that have stopped growing. A fully grown plant cell may retain its primary cell wall (sometimes thickening it), or it may deposit an additional, rigidifying layer of different composition; this is the secondary wall. Secondary cell walls are responsible for most of the plant's mechanical support as well as the mechanical properties prized in wood. In contrast to the permanent stiffness and load-bearing capacity of thick secondary walls, the thin primary walls are capable of serving a structural, supportive role only when the vacuoles within the cell are filled with water to the point that they exert a turgor pressure against the cell wall. Turgor-induced stiffening of primary walls is analogous to the stiffening of the sides of a pneumatic tire by air pressure. The wilting of flowers and leaves is caused by a loss of turgor pressure, which results in turn from the loss of water from the plant cells.

Components of the Cell Wall

Although primary and secondary wall layers differ in detailed chemical composition and structural organization, their basic architecture is the same, consisting of cellulose fibres of great tensile strength embedded in a water-saturated matrix of polysaccharides and structural glycoproteins.

Cellulose

Cellulose consists of several thousand glucose molecules linked end to end. The chemical links between the individual glucose subunits give each cellulose molecule a flat ribbonlike structure that allows adjacent molecules to band laterally together into microfibrils with lengths ranging from two to seven micrometres. Cellulose fibrils are synthesized by enzymes floating in the cell membrane and are arranged in a rosette configuration. Each rosette appears capable of "spinning" a microfibril into the cell wall. During this process, as new glucose subunits are added to the growing end of the fibril, the rosette is pushed around the cell on the surface of the cell membrane, and its cellulose fibril becomes wrapped around the protoplast. Thus, each plant cell can be viewed as making its own cellulose fibril cocoon.

Matrix polysaccharides

The two major classes of cell wall matrix polysaccharides are the hemicelluloses and the pectic polysaccharides, or pectins. Both are synthesized in the Golgi apparatus, brought to the cell surface in small vesicles, and secreted into the cell wall.

Hemicelluloses consist of glucose molecules arranged end to end as in cellulose, with short side chains of xylose and other uncharged sugars attached to one side of the ribbon. The other side of the ribbon binds tightly to the surface of cellulose fibrils, thereby coating the microfibrils with hemicellulose and preventing them from adhering together in an uncontrolled manner. Hemicellulose molecules have been shown to regulate the rate at which primary cell walls expand during growth.

The heterogeneous, branched, and highly hydrated pectic polysaccharides differ from hemicelluloses in important respects. Most notably, they are negatively charged because of galacturonic acid residues, which, together with rhamnose sugar molecules, form the linear backbone of all pectic polysaccharides. The backbone contains stretches of pure galacturonic acid residues interrupted by segments in which galacturonic acid and rhamnose residues alternate; attached to these latter segments are complex, branched sugar side chains. Because of their negative charge, pectic polysaccharides bind tightly to positively charged ions, or cations. In cell walls, calcium ions cross-link the stretches of pure galacturonic acid residues tightly, while leaving the rhamnose-containing segments in a more open, porous configuration. This cross-linking creates the semirigid gel properties characteristic of the cell wall matrix—a process exploited in the preparation of jellied preserves.

Proteins

Although plant cell walls contain only small amounts of protein, they serve a number of important functions. The most prominent group are the hydroxyproline-rich glycoproteins, shaped like rods with connector sites, of which extensin is a prominent example. Extensin contains 45 percent hydroxyproline and 14 percent serine residues distributed along its length. Every hydroxyproline residue carries a short side chain of arabinose sugars, and most serine residues carry a galactose sugar; this gives rise to long molecules, resembling bottle brushes, that are secreted into the cell wall toward the end of primary-wall formation and become covalently cross-linked into a mesh at the time that cell growth stops. Plant cells may control their ultimate size by regulating the time at which this cross-linking of extensin molecules occurs.

In addition to the structural proteins, cell walls contain a variety of enzymes. Most notable are those that cross-link extensin, lignin, cutin, and suberin molecules into networks. Other enzymes help protect plants against fungal pathogens by breaking fragments off of the cell walls of the fungi. The fragments in turn induce defense responses in underlying cells. The softening of ripe fruit and dropping of leaves in the autumn are brought about by cell wall-degrading enzymes.

Plastics
Cell wall plastics such as lignin, cutin, and suberin all contain a variety of organic compounds cross-linked into tight three-dimensional networks that strengthen cell walls and make them more resistant to fungal and bacterial attack. Lignin is the general name for a diverse group of polymers of aromatic alcohols. Deposited mostly in secondary cell walls and providing the rigidity of terrestrial vascular plants, it accounts for up to 30 percent of a plant's dry weight. The diversity of cross-links between the polymers—and the resulting tightness—makes lignin a formidable barrier to the penetration of most microbes.

Agave shawii (top) and Echeveria (bottom), two types of xerophytes (plants adapted to arid habitats). They develop highly cutinized fleshy leaves and stems for water storage with which they modulate the effects of strong sunlight, low humidity, and scant water.

Cutin and suberin are complex biopolyesters composed of fatty acids and aromatic compounds. Cutin is the major component of the cuticle, the waxy, water-repelling surface layer of cell walls exposed to the environment aboveground. By reducing the wetability of leaves and stems—and thereby affecting the ability of fungal spores to germinate—it plays an important part in the defense strategy of plants. Suberin serves with waxes as a surface barrier of underground parts. Its synthesis is also stimulated in cells close to wounds, thereby sealing off the wound surfaces and protecting underlying cells from dehydration.

Intercellular Communication
Plasmodesmata

Similar to the gap junction of animal cells is the plasmodesma, a channel passing through the cell wall and allowing direct molecular communication between adjacent plant cells. Plasmodesmata are lined with cell membrane, in effect uniting all connected cells with one continuous cell membrane. Running down the middle of each channel is a thin membranous tube that connects the endoplasmic reticula (ER) of the two cells. This structure is a remnant of the ER of the original parent cell, which, as the parent cell divided, was caught in the developing cell plate.

Although the precise mechanisms are not fully understood, the plasmodesma is thought to regulate the passage of small molecules such as salts, sugars, and amino acids by constricting or dilating the openings at each end of the channel.

Oligosaccharides With Regulatory Functions

The discovery of cell wall fragments with regulatory functions opened a new era in plant research. For years scientists had been puzzled by the chemical complexity of cell wall polysaccharides, which far exceeds the structural requirements of plant cell walls. The answer came when it was found that specific fragments of cell wall polysaccharides, called oligosaccharins, are able to induce specific responses in plant cells and tissues. One such fragment, released by enzymes used by fungi to break down plant cell walls, consists of a linear polymer of 10 to 12 galacturonic acid residues. Exposure of plant cells to such fragments induces them to produce antibiotics known as phytoalexins. In other experiments it has been shown that exposing strips of tobacco stem cells to a different type of cell wall fragment leads to the growth of roots; other fragments lead to the formation of stems, and yet others to the production of flowers. In all instances the concentration of oligosaccharins required to bring about the observed responses is equal to that of hormones in animal cells; indeed, oligosaccharins may be viewed as the oligosaccharide hormones of plants.

Cell Division And Growth

In unicellular organisms, cell division is the means of reproduction; in multicellular organisms, it is the means of tissue growth and maintenance. Survival of the eukaryotes depends upon interactions between many cell types, and it is essential that a balanced distribution of types be maintained. This is achieved by the highly regulated process of cell proliferation. The growth and division of different cell populations are regulated in different ways, but the basic mechanisms are similar throughout multicellular organisms.

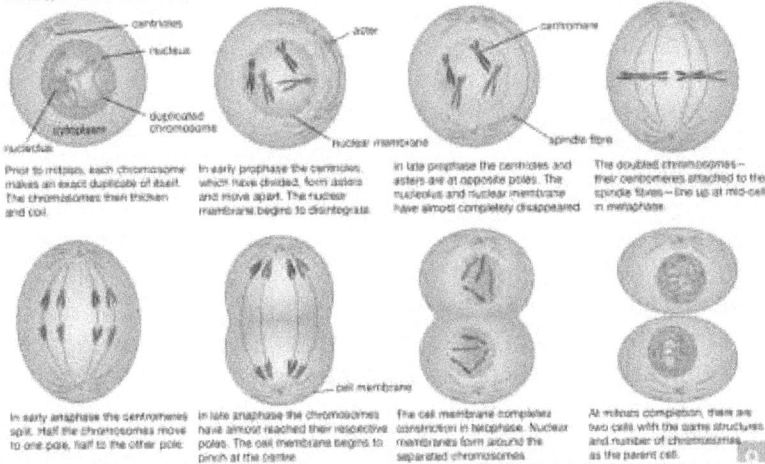

Mitosis

One cell gives rise to two genetically identical daughter cells during the process of mitosis.

Most tissues of the body grow by increasing their cell number, but this growth is highly regulated to maintain a balance between different tissues. In adults most cell division is involved in tissue renewal rather than growth, many types of cells undergoing continuous replacement. Skin cells, for example, are constantly being sloughed off and replaced; in this case, the mature differentiated cells do not divide, but their population is renewed by division of immature stem cells. In certain other cells, such as those of the liver, mature cells remain capable of division to allow growth or regeneration after injury.

In contrast to these patterns, other types of cells either cannot divide or are prevented from dividing by certain molecules produced by nearby cells. As a result, in the adult organism, some tissues have a greatly reduced capacity to renew damaged or diseased cells. Examples of such tissues include heart muscle, nerve cells of the central nervous system, and lens cells in mammals. Maintenance and repair of these cells is limited to replacing intracellular components rather than replacing entire cells.

Duplication of the Genetic Material

Before a cell can divide, it must accurately and completely duplicate the genetic information encoded in its DNA in order for its progeny cells to function and survive. This is a complex problem because of the great length of DNA molecules. Each human chromosome consists of a long double spiral, or helix, each strand of which consists of more than 100 million nucleotides.

The duplication of DNA is called DNA replication, and it is initiated by complex enzymes called DNA polymerases. These progress along the molecule, reading the sequences of nucleotides that are linked together to make DNA chains. Each strand of the DNA double helix, therefore, acts as a template specifying the nucleotide structure of a new growing chain. After replication, each of the two daughter DNA double helices consists of one parental DNA strand wound around one newly synthesized DNA strand.

In order for DNA to replicate, the two strands must be unwound from each other. Enzymes called helicases unwind the two DNA strands, and additional proteins bind to the separated strands to stabilize them and prevent them from pairing again. In addition, a remarkable class of enzyme called DNA topoisomerase removes the helical twists by cutting either one or both strands and then resealing the cut. These enzymes can also untangle and unknot DNA when it is tightly coiled into a chromatin fibre.

In the circular DNA of prokaryotes, replication starts at a unique site called the origin of replication and then proceeds in both directions around the molecule until the two processes meet, producing two daughter molecules. In rapidly growing prokaryotes, a second round of replication can start before the first has finished. The situation in eukaryotes is more complicated, as replication moves more slowly than in prokaryotes. At 500 to 5,000 nucleotides per minute (versus 100,000 nucleotides per minute in prokaryotes), it would take a human chromosome about a month to replicate if started at a single site. Actually, replication begins at many sites on the long chromosomes of animals, plants, and fungi. Distances between adjacent initiation sites are not always the same; for example, they are closer in the rapidly dividing embryonic cells of frogs or flies than in adult cells of the same species.

Accurate DNA replication is crucial to ensure that daughter cells have exact copies of the genetic information for synthesizing proteins. Accuracy is achieved by a "proofreading" ability of the DNA polymerase itself. It can erase its own errors and then synthesize anew. There are also repair systems that correct genetic damage to DNA. For example, the incorporation of an incorrect nucleotide, or damage caused by mutagenic agents, can be corrected by cutting out a section of the daughter strand and recopying the parental strand.

Cell Division
Mitosis and cytokinesis

In eukaryotes the processes of DNA replication and cell division occur at different times of the cell division cycle. During cell division, DNA condenses to form short, tightly coiled, rodlike chromosomes. Each chromosome then splits longitudinally, forming two identical chromatids. Each pair of chromatids is divided between the two daughter cells during mitosis, or division of the nucleus, a process in which the chromosomes are propelled by attachment to a bundle of microtubules called the mitotic spindle.

Mitosis can be divided into five phases. In prophase the mitotic spindle forms and the chromosomes condense. In prometaphase the nuclear envelope breaks down (in many but not all eukaryotes) and the chromosomes attach to the mitotic spindle. Both chromatids of each chromosome attach to the spindle at a specialized chromosomal region called the kinetochore. In metaphase the condensed chromosomes align in a plane across the equator of the mitotic spindle. Anaphase follows as the separated chromatids move abruptly toward opposite spindle poles. Finally, in telophase a new nuclear envelope forms around each set of unraveling chromatids.

An essential feature of mitosis is the attachment of the chromatids to opposite poles of the mitotic spindle. This ensures that each of the daughter cells will receive a complete set of chromosomes. The mitotic spindle is composed of microtubules, each of which is a tubular assembly of molecules of the protein tubulin. Some microtubules extend from one spindle pole to the other, while a second class extends from one spindle pole to a chromatid. Microtubules can grow or shrink by the addition or removal of tubulin molecules. The shortening of spindle microtubules at anaphase propels attached chromatids to the spindle poles, where they unravel to form new nuclei.

The two poles of the mitotic spindle are occupied by centrosomes, which organize the microtubule arrays. In animal cells each centrosome contains a pair of cylindrical centrioles, which are themselves composed of complex arrays of microtubules. Centrioles duplicate at a precise time in the cell division cycle, usually close to the start of DNA replication.

After mitosis comes cytokinesis, the division of the cytoplasm. This is another process in which animal and plant cells differ. In animal cells cytokinesis is achieved through the constriction of the cell by a ring of contractile microfilaments consisting of actin and myosin, the proteins involved in muscle contraction and other forms of cell movement. In plant cells the cytoplasm is divided by the formation of a new cell wall, called the cell plate, between the two daughter cells. The cell plate arises from small Golgi-derived vesicles that coalesce in a plane across the equator of the late telophase spindle to form a disk-shaped structure. In this process, each vesicle contributes its membrane to the forming cell membranes and its matrix contents to the forming cell wall. A second set of vesicles extends the edge of the cell plate until it reaches and fuses with the sides of the parent cell, thereby completely separating the two new daughter cells. At this point, cellulose synthesis commences, and the cell plate becomes a primary cell wall .

Meiosis

A specialized division of chromosomes called meiosis occurs during the formation of the reproductive cells, or gametes, of sexually reproducing organisms. Gametes such as ova, sperm, and pollen begin as germ cells, which, like other types of cells, have two copies of each gene in their nuclei. The chromosomes composed of these matching genes are called homologs. During DNA replication, each chromosome duplicates into two attached chromatids. The homologous chromosomes are then separated to opposite poles of the meiotic spindle by microtubules similar to those of the mitotic spindle. At this stage in the meiosis of germ cells, there is a crucial difference from the mitosis of other cells. In meiosis the two chromatids making up each chromosome remain together, so that whole chromosomes are separated from their homologous partners. Cell division then occurs, followed by a second division that resembles mitosis more closely in that it separates the two chromatids of each remaining chromosome. In this way, when meiosis is complete, each mature gamete receives only one copy of each gene instead of the two copies present in other cells.

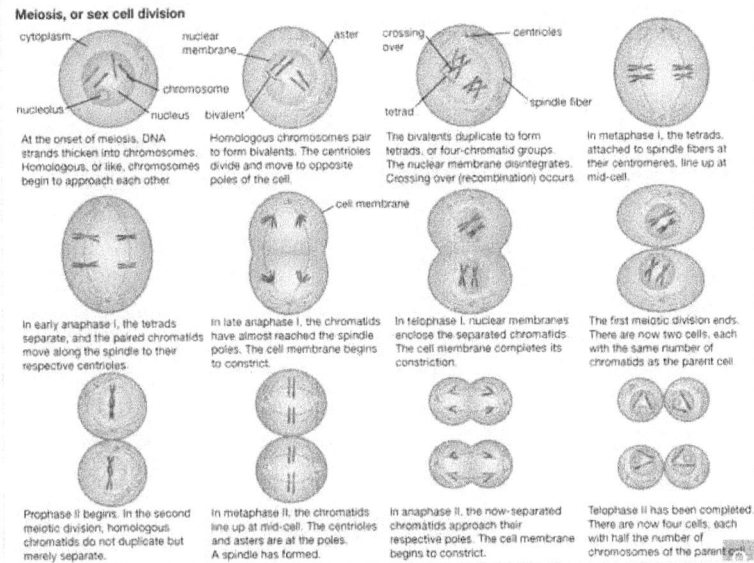

The formation of gametes (sex cells) occurs during the process of meiosis.

The cell division cycle

In prokaryotes, DNA synthesis can take place uninterrupted between cell divisions, and new cycles of DNA synthesis can begin before previous cycles have finished. In contrast, eukaryotes duplicate their DNA exactly once during a discrete period between cell divisions. This period is called the S (for synthetic) phase. It is preceded by a period called G1 (meaning "first gap") and followed by a period called G2, during which nuclear DNA synthesis does not occur.

The four periods G1, S, G2, and M (for mitosis) make up the cell division cycle. The cell cycle characteristically lasts between 10 and 20 hours in rapidly proliferating adult cells, but it can be arrested for weeks or months in quiescent cells or for a lifetime in neurons of the brain. Prolonged arrest of this type usually occurs during the G1 phase and is sometimes referred to as G0. In contrast, some embryonic cells, such as those of fruit flies (vinegar flies), can complete entire cycles and divide in only 11 minutes. In these exceptional cases, G1 and G2 are undetectable, and mitosis alternates with DNA synthesis. In addition, the duration of the S phase varies dramatically. The fruit fly embryo takes only four minutes to replicate its DNA, compared with several hours in adult cells of the same species.

Controlled Proliferation

Several studies have identified the transition from the G1 to the S phase as a crucial control point of the cell cycle. Stimuli are known to cause resting cells to proliferate by inducing them to leave G1 and begin DNA synthesis. These stimuli, called growth factors, are naturally occurring proteins specific to certain groups of cells in the body. They include nerve growth factor, epidermal growth factor, and platelet-derived growth factor. Such factors may have important roles in the healing of wounds as well as in the maintenance and growth of normal tissues. Many growth factors are known to act on the external membrane of the cell, by interacting with specialized protein receptor molecules. These respond by triggering further cellular changes, including an increase in calcium levels that makes the cell interior more alkaline and the addition of phosphate groups to the amino acid tyrosine in proteins. The complex response of cells to growth factors is of fundamental importance to the control of cell proliferation.

Failure of Proliferation Control

Cancer can arise when the controlling factors over cell growth fail and allow a cell and its descendants to keep dividing at the expense of the organism. Studies of viruses that transform cultured cells and thus lead to the loss of control of cell growth have provided insight into the mechanisms that drive the formation of tumours. Transformed cells may differ from their normal progenitors by continuing to proliferate at very high densities, in the absence of growth factors, or in the absence of a solid substrate for support.

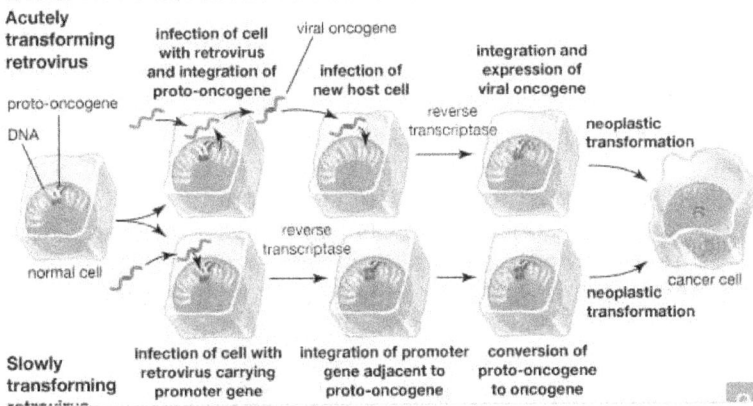

Cancer-Causing Retroviruses

Retroviral insertion can convert a proto-oncogene, integral to the control of cell division, into an oncogene, the agent responsible for transforming a healthy cell into a cancer cell. An acutely transforming retrovirus (shown at top), which produces tumours within weeks of infection, incorporates genetic material from a host cell into its own genome upon infection, forming a viral oncogene. When the viral oncogene infects another cell, an enzyme called reverse transcriptase copies the single-stranded genetic material into double-stranded DNA, which is then integrated into the cellular genome. A slowly transforming retrovirus (shown at bottom), which requires months to elicit tumour growth, does not disrupt cellular function through the insertion of a viral oncogene. Rather, it carries a promoter gene that is integrated into the cellular genome of the host cell next to or within a proto-oncogene, allowing conversion of the proto-oncogene to an oncogene.

Major advances in the understanding of growth control have come from studies of the viral genes that cause transformation. These viral oncogenes have led to the identification of related cellular genes called protooncogenes. Protooncogenes can be altered by mutation or epigenetic modification, which converts them into oncogenes and leads to cell transformation. Specific oncogenes are activated in particular human cancers. For example, an oncogene called RAS is associated with many epithelial cancers, while another, called MYC, is associated with leukemias.

An interesting feature of oncogenes is that they may act at different levels corresponding to the multiple steps seen in the development of cancer. Some oncogenes immortalize cells so that they divide indefinitely, whereas normal cells die after a limited number of generations. Other oncogenes transform cells so that they grow in the absence of growth factors. A combination of these two functions leads to loss of proliferation control, whereas each of these functions on its own cannot. The mode of action of oncogenes also provides important clues to the nature of growth control and cancer. For example, some oncogenes are known to encode receptors for growth factors that may cause continuous proliferation in the absence of appropriate growth factors.

Loss of growth control has the added consequence that cells no longer repair their DNA effectively, and thus aberrant mitoses occur. As a result, additional mutations arise that subvert a cell's normal constraints to remain in its tissue of origin. Epithelial tumour cells, for example, acquire the ability to cross the basal lamina and enter the bloodstream or lymphatic system, where they migrate to other parts of the body, a process called metastasis. When cells metastasize to distant tissues, the tumour is described as malignant, whereas prior to metastasis a tumour is described as benign.

Cell Differentiation

Adult organisms are composed of a number of distinct cell types. Cells are organized into tissues, each of which typically contains a small number of cell types and is devoted to a specific physiological function. For example, the epithelial tissue lining the small intestine contains columnar absorptive cells, mucus-secreting goblet cells, hormone-secreting endocrine cells, and enzyme-secreting Paneth cells. In addition, there exist undifferentiated dividing cells that lie in the crypts between the intestinal villi and serve to replace the other cell types when they become damaged or worn out. Another example of a differentiated tissue is the skeletal tissue of a long bone, which contains osteoblasts (large cells that synthesize bone) in the outer sheath and osteocytes (mature bone cells) and osteoclasts (multinucleate cells involved in bone remodeling) within the matrix.

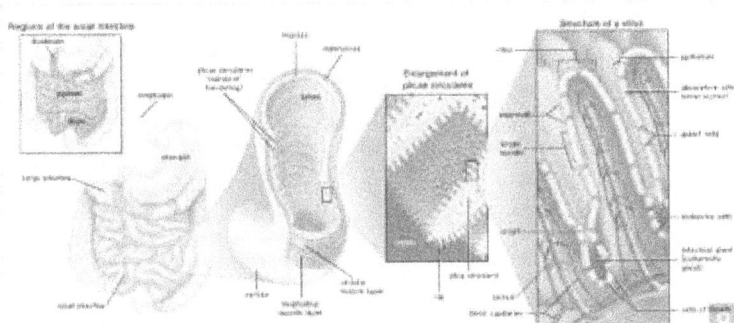

The small intestine contains many distinct types of cells, each of which serves a specific function.

In general, the simpler the overall organization of the animal, the fewer the number of distinct cell types that they possess. Mammals contain more than 200 different cell types, whereas simple invertebrate animals may have only a few different types. Plants are also made up of differentiated cells, but they are quite different from the cells of animals. For example, a leaf in a higher plant is covered with a cuticle layer of epidermal cells. Among these are pores composed of two specialized cells, which regulate gaseous exchange across the epidermis. Within the leaf is the mesophyll, a spongy tissue responsible for photosynthetic activity. There are also veins composed of xylem elements, which transport water up from the soil, and phloem elements, which transport products of photosynthesis to the storage organs.

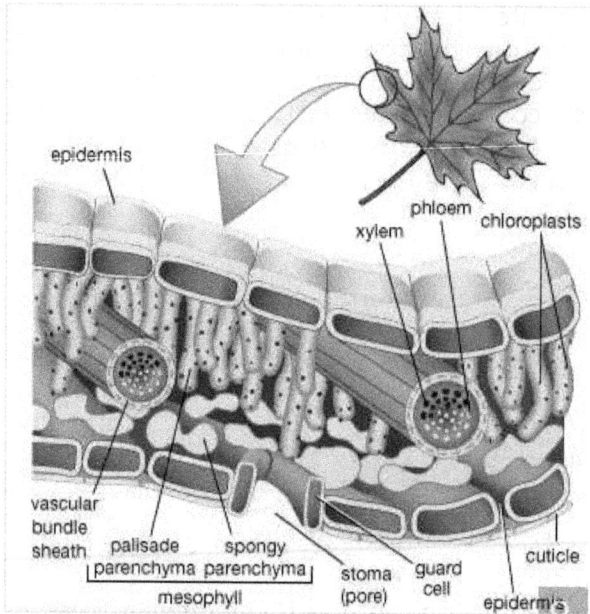

Structures of a leafThe epidermis is often covered with a waxy protective cuticle that helps prevent water loss from inside the leaf. Oxygen, carbon dioxide, and water enter and exit the leaf through pores (stomata) scattered mostly along the lower epidermis. The stomata are opened and closed by the contraction and expansion of surrounding guard cells. The vascular, or conducting, tissues are known as xylem and phloem; water and minerals travel up to the leaves from the roots through the xylem, and sugars made by photosynthesis are transported to other parts of the plant through the phloem. Photosynthesis occurs within the chloroplast-containing mesophyll layer.

The various cell types have traditionally been recognized and classified according to their appearance in the light microscope following the process of fixing, processing, sectioning, and staining tissues that is known as histology. Classical histology has been augmented by a variety of more discriminating techniques. Electron microscopy allows for higher magnifications. Histochemistry involves the use of coloured precipitating substrates to stain particular enzymes in situ. Immunohistochemistry uses specific antibodies to identify particular substances, usually proteins or carbohydrates, within cells. In situ hybridization involves the use of nucleic acid probes to visualize the location of specific messenger RNAs (mRNA). These modern methods have allowed the identification of more cell types than could be visualized by classical histology, particularly in the brain, the immune system, and among the hormone-secreting cells of the endocrine system.

The Differentiated State

The biochemical basis of cell differentiation is the synthesis by the cell of a particular set of proteins, carbohydrates, and lipids. This synthesis is catalyzed by proteins called enzymes. Each enzyme in turn is synthesized in accordance with a particular gene, or sequence of nucleotides in the DNA of the cell nucleus. A particular state of differentiation, then, corresponds to the set of genes that is expressed and the level to which it is expressed.

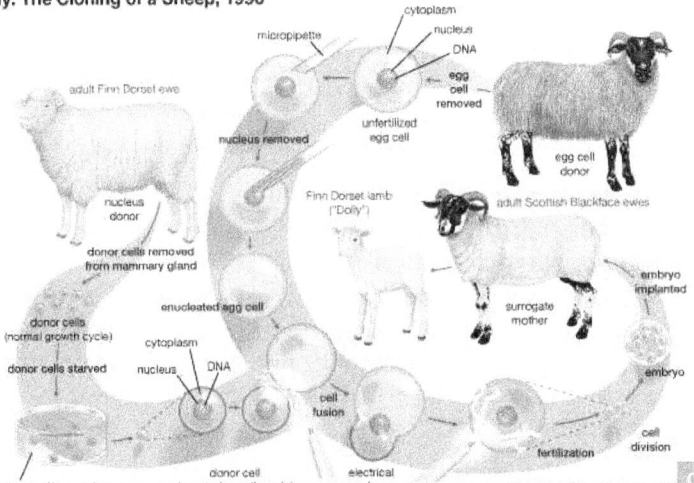

Dolly: The Cloning of a Sheep, 1996

Dolly the sheep was successfully cloned in 1996 by fusing the nucleus from a mammary-gland cell of a Finn Dorset ewe into an enucleated egg cell taken from a Scottish Blackface ewe. Carried to term in the womb of another Scottish Blackface ewe, Dolly was a genetic copy of the Finn Dorset ewe.

It is believed that all of an organism's genes are present in each cell nucleus, no matter what the cell type, and that differences between tissues are not due to the presence or absence of certain genes but are due to the expression of some and the repression of others. In animals the best evidence for retention of the entire set of genes comes from whole animal cloning experiments in which the nucleus of a differentiated cell is substituted for the nucleus of a fertilized egg. In many species this can result in the development of a normal embryo that contains the full range of body parts and cell types. Likewise, in plants it is often possible to grow complete embryos from individual cells in tissue culture. Such experiments show that any nucleus has the genetic information required for the growth of a developing organism, and they strongly suggest that, for most tissues, cell differentiation arises from the regulation of genetic activity rather than the removal or destruction of unwanted genes. The only known exception to this rule comes from the immune system, where segments of DNA in developing white blood cells are slightly rearranged, producing a wide variety of antibody and receptor molecules.

At the molecular level there are many ways in which the expression of a gene can be differentially regulated in different cell types. There may be differences in the copying, or transcription, of the gene into RNA; in the processing of the initial RNA transcript into mRNA; in the control of mRNA movement to the cytoplasm; in the translation of mRNA to protein; or in the stability of mRNA. However, the control of transcription has the most influence over gene expression and has received the most detailed analysis.

The DNA in the cell nucleus exists in the form of chromatin, which is made up of DNA bound to histones (simple alkaline proteins) and other nonhistone proteins. Most of the DNA is complexed into repeating structures called nucleosomes, each of which contains eight molecules of histone. Active genes are found in parts of the DNA where the chromatin has an "open" configuration, in which regulatory proteins are able to gain access to the DNA. The degree to which the chromatin opens depends on chemical modifications of the outer parts of the histone molecules and on the presence or absence of particular nonhistone proteins. Transcriptional control is exerted with the help of regulatory sequences that are found associated with a gene, such as the promoter sequence, a region near the start of the gene, and enhancer sequences, regions that lie elsewhere within the DNA that augment the activity of enzymes involved in the process of transcription. Whether or not transcription occurs depends on the binding of transcription factors to these regulatory sequences. Transcription factors are proteins that usually possess a DNA-binding region, which recognizes the specific regulatory sequence in the DNA, and an effector region, which activates or inhibits transcription. Transcription factors often work by recruiting enzymes that add modifications (e.g., acetyl groups or methyl groups) to or remove modifications from the outer parts of the histone molecules. This controls the folding of the chromatin and the accessibility of the DNA to RNA polymerase and other transcription factors.

In general, it requires several transcription factors working in combination to activate a gene. For example, the chicken delta 1 crystallin gene, normally expressed only in the lens of the eye, has a promoter that contains binding sites for two activating transcription factors and an enhancer that contains binding sites for two other activating transcription factors. There is also an additional enhancer site that can bind either an activator (deltaEF3) or a repressor (deltaEF1). Successful transcription requires that all these sites are occupied by the correct transcription factors.

Fully differentiated cells are qualitatively different from one another. States of terminal differentiation are stable and persistent, both in the lifetime of the cell and in successive cell generations (in the case of differentiated types that are capable of continued cell division). The inherent stability of the differentiated state is maintained by various processes, including feedback activation of genes by their own products and repression of inactive genes. Chromatin structure may be important in maintaining states of differentiation, although it is still unclear whether this can be maintained during DNA replication, which involves temporary removal of chromosomal proteins and unwinding of the DNA double helix.

A type of differentiation control that is maintained during DNA replication is the methylation of DNA, which tends to recruit histone deacetylases and hence close up the structure of the chromatin. DNA methylation occurs when a methyl group is attached to the exterior, or sugar-phosphate side, of a cytosine (C) residue. Cytosine methylation occurs only on a C nucleotide when it is connected to a G (guanine) nucleotide on the same strand of DNA. These nucleotide pairings are called CG dinucleotides. One class of DNA methylase enzyme can introduce new methylations when required, whereas another class, called maintenance methylases, methylates CG dinucleotides in the DNA double helix only when the CG of the complementary strand is already methylated. Each time the methylated DNA is replicated, the old strand has the methyl groups and the new strand does not. The maintenance methylase will then add methyl groups to all the CGs opposite the existing methyl groups to restore a fully methylated double helix. This mechanism guarantees stability of the DNA methylation pattern, and hence the differentiated state, during the processes of DNA replication and cell division.

The Process of Differentiation

Differentiation from visibly undifferentiated precursor cells occurs during embryonic development, during metamorphosis of larval forms, and following the separation of parts in asexual reproduction. It also takes place in adult organisms during the renewal of tissues and the regeneration of missing parts. Thus, cell differentiation is an essential and ongoing process at all stages of life.

The visible differentiation of cells is only the last of a progressive sequence of states. In each state, the cell becomes increasingly committed toward one type of cell into which it can develop. States of commitment are sometimes described as "specification" to represent a reversible type of commitment and as "determination" to represent an irreversible commitment. Although states of specification and determination both represent differential gene activity, the properties of embryonic cells are not necessarily the same as those of fully differentiated cells. In particular, cells in specification states are usually not stable over prolonged periods of time.

Two mechanisms bring about altered commitments in the different regions of the early embryo: cytoplasmic localization and induction. Cytoplasmic localization is evident in the earliest stages of development of the embryo. During this time, the embryo divides without growth, undergoing cleavage divisions that produce separate cells called blastomeres. Each blastomere inherits a certain region of the original egg cytoplasm, which may contain one or more regulatory substances called cytoplasmic determinants. When the embryo has become a solid mass of blastomeres (called a morula), it generally consists of two or more differently committed cell populations—a result of the blastomeres having incorporated different cytoplasmic determinants. Cytoplasmic determinants may consist of mRNA or protein in a particular state of activation. An example of the influence of a cytoplasmic determinant is a receptor called Toll, located in the membranes of Drosophila (fruit fly) eggs. Activation of Toll ensures that the blastomeres will develop into ventral (underside) structures, while blastomeres containing inactive Toll will produce cells that will develop into dorsal (back) structures.

In induction, the second mechanism of commitment, a substance secreted by one group of cells alters the development of another group. In early development, induction is usually instructive; that is, the tissue assumes a different state of commitment in the presence of the signal than it would in the absence of the signal. Inductive signals often take the form of concentration gradients of substances that evoke a number of different responses at different concentrations. This leads to the formation of a sequence of groups of cells, each in a different state of specification. For example, in Xenopus (clawed frog) the early embryo contains a signaling centre called the organizer that secretes inhibitors of bone morphogenetic proteins (BMPs), leading to a ventral-to-dorsal (belly-to-back) gradient of BMP activity. The activity of BMP in the ventral region of the embryo suppresses the expression of transcription factors involved in the formation of the central nervous system and segmented muscles. Suppression ensures that these structures are formed only on the dorsal side, where there is decreased activity of BMP.

The final stage of differentiation often involves the formation of several types of differentiated cells from one precursor or stem cell population. Terminal differentiation occurs not only in embryonic development but also in many tissues in postnatal life. Control of this process depends on a system of lateral inhibition in which cells that are differentiating along a particular pathway send out signals that repress similar differentiation by their neighbours. For example, in the developing central nervous system of vertebrates, neurons arise from a simple tube of neuroepithelium, the cells of which possess a surface receptor called Notch. These cells also possess another cell surface molecule called Delta that can bind to and activate Notch on adjacent cells. Activation of Notch initiates a cascade of intracellular events that results in suppression of Delta production and suppression of neuronal differentiation. This means that the neuroepithelium generates only a few cells with high expression of Delta surrounded by a larger number of cells with low expression of Delta. High Delta production and low Notch activation makes the cells develop into neurons. Low Delta production and high Notch activation makes the cells remain as precursor cells or become glial (supporting) cells. A similar mechanism is known to produce the endocrine cells of the pancreas and the goblet cells of the intestinal epithelium. Such lateral inhibition systems work because cells in a population are never quite identical to begin with. There are always small differences, such as in the number of Delta molecules displayed on the cell surface. The mechanism of lateral inhibition amplifies these small differences, using them to bring about differential gene expression that leads to stable and persistent states of cell differentiation.

Errors in Differentiation

Three classes of abnormal cell differentiation are dysplasia, metaplasia, and anaplasia. Dysplasia indicates an abnormal arrangement of cells, usually arising from a disturbance in their normal growth behaviour. Some dysplasias are precursor lesions to cancer, whereas others are harmless and regress spontaneously. For example, dysplasia of the uterine cervix, called cervical intraepithelial neoplasia (CIN), may progress to cervical cancer. It can be detected by cervical smear cytology tests (Pap smears).

Metaplasia is the conversion of one cell type into another. In fact, it is not usually the differentiated cells themselves that change but rather the stem cell population from which they are derived. Metaplasia commonly occurs where chronic tissue damage is followed by extensive regeneration. For example, squamous metaplasia of the bronchi occurs when the ciliated respiratory epithelial cells of people who smoke develop into squamous, or flattened, cells. In intestinal metaplasia of the stomach, patches resembling intestinal tissue arise in the gastric mucosa, often in association with gastric ulcers. Both of these types of metaplasia may progress to cancer.

Anaplasia is a loss of visible differentiation that can occur in advanced cancer. In general, early cancers resemble their tissue of origin and are described and classified by their pattern of differentiation. However, as they develop, they produce variants of more abnormal appearance and increased malignancy. Finally, a highly anaplastic growth can occur, in which the cancerous cells bear no visible relation to the parent tissue.

The Evolution of Cells
The development of genetic information

Life on Earth could not exist until a collection of catalysts appeared that could promote the synthesis of more catalysts of the same kind. Early stages in the evolutionary pathway of cells presumably centred on RNA molecules, which not only present specific catalytic surfaces but also contain the potential for their own duplication through the formation of a complementary RNA molecule. It is assumed that a small RNA molecule eventually appeared that was able to catalyze its own duplication.

Imperfections in primitive RNA replication likely gave rise to many variant autocatalytic RNA molecules. Molecules of RNA that acquired variations that increased the speed or the fidelity of self-replication would have outmultiplied other, less-competent RNA molecules. In addition, other small RNA molecules that existed in symbiosis with autocatalytic RNA molecules underwent natural selection for their ability to catalyze useful secondary reactions such as the production of better precursor molecules. In this way, sophisticated families of RNA catalysts could have evolved together, since cooperation between different molecules produced a system that was much more effective at self-replication than a collection of individual RNA catalysts.

Another major step in the evolution of the cell would have been the development, in one family of self-replicating RNA, of a primitive mechanism of protein synthesis. Protein molecules cannot provide the information for the synthesis of other protein molecules like themselves. This information must ultimately be derived from a nucleic acid sequence. Protein synthesis is much more complex than RNA synthesis, and it could not have arisen before a group of powerful RNA catalysts evolved. Each of these catalysts presumably has its counterpart among the RNA molecules that function in the current cell: (1) there was an information RNA molecule, much like messenger RNA (mRNA), whose nucleotide sequence was read to create an amino acid sequence; (2) there was a group of adaptor RNA molecules, much like transfer RNA (tRNA), that could bind to both mRNA and a specific activated amino acid; and (3) finally, there was an RNA catalyst, much like ribosomal RNA (rRNA), that facilitated the joining together of the amino acids aligned on the mRNA by the adaptor RNA.

At some point in the evolution of biological catalysts, the first cell was formed. This would have required the partitioning of the primitive biological catalysts into individual units, each surrounded by a membrane. Membrane formation might have occurred quite simply, since many amphiphilic molecules—half hydrophobic (water-repelling) and half hydrophilic (water-loving)—aggregate to form bilayer sheets in which the hydrophobic portions of the molecules line up in rows to form the interior of the sheet and leave the hydrophilic portions to face the water. Such bilayer sheets can spontaneously close up to form the walls of small, spherical vesicles, as can the phospholipid bilayer membranes of present-day cells.

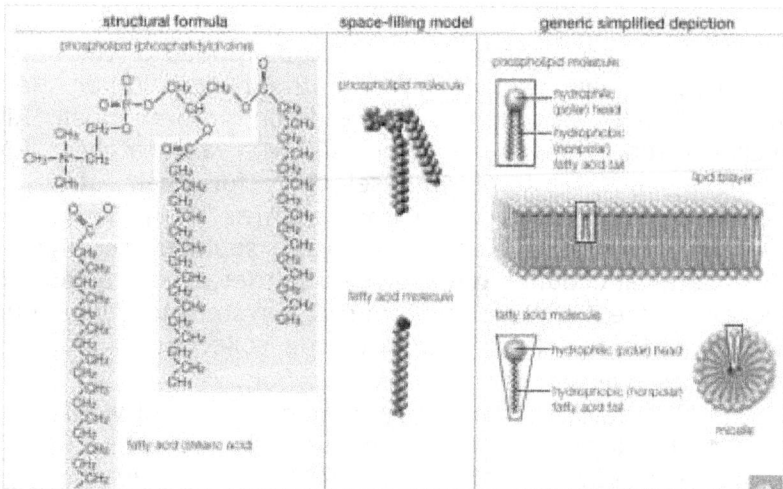

Structure and properties of two representative lipidsBoth stearic acid (a fatty acid) and phosphatidylcholine (a phospholipid) are composed of chemical groups that form polar "heads" and nonpolar "tails." The polar heads are hydrophilic, or soluble in water, whereas the nonpolar tails are hydrophobic, or insoluble in water. Lipid molecules of this composition spontaneously form aggregate structures such as micelles and lipid bilayers, with their hydrophilic ends oriented toward the watery medium and their hydrophobic ends shielded from the water.

As soon as the biological catalysts became compartmentalized into small individual units, or cells, the units would have begun to compete with one another for the same resources. The active competition that ensued must have greatly accelerated evolutionary change, serving as a powerful force for the development of more efficient cells. In this way, cells eventually arose that contained new catalysts, enabling them to use simpler, more abundant precursor molecules for their growth. Because these cells were no longer dependent on preformed ingredients for their survival, they were able to spread far beyond the limited environments where the first primitive cells arose.

It is often assumed that the first cells appeared only after the development of a primitive form of protein synthesis. However, it is by no means certain that cells cannot exist without proteins, and it has been suggested that the first cells contained only RNA catalysts. In either case, protein molecules, with their chemically varied side chains, are more powerful catalysts than RNA molecules; therefore, as time passed, cells arose in which RNA served primarily as genetic material, being directly replicated in each generation and inherited by all progeny cells in order to specify proteins.

As cells became more complex, a need would have arisen for a stabler form of genetic information storage than that provided by RNA. DNA, related to RNA yet chemically stabler, probably appeared rather late in the evolutionary history of cells. Over a period of time, the genetic information in RNA sequences was transferred to DNA sequences, and the ability of RNA molecules to replicate directly was lost. It was only at this point that the central process of biology—the synthesis, one after the other, of DNA, RNA, and protein—appeared.

The Development of Metabolism

The first cells presumably resembled prokaryotic cells in lacking nuclei and functional internal compartments, or organelles. These early cells were also anaerobic (not requiring oxygen), deriving their energy from the fermentation of organic molecules that had previously accumulated on the Earth over long periods of time. Eventually, more sophisticated cells evolved that could carry out primitive forms of photosynthesis, in which light energy was harnessed by membrane-bound proteins to form organic molecules with energy-rich chemical bonds. A major turning point in the evolution of life was the development of photosynthesizing prokaryotes requiring only water as an electron donor and capable of producing molecular oxygen. The descendants of these prokaryotes, the blue-green algae (cyanobacteria), still exist as viable life-forms. Their ancestors prospered to such an extent that the atmosphere became rich in the oxygen they produced. The free availability of this oxygen in turn enabled other prokaryotes to evolve aerobic forms of metabolism that were much more efficient in the use of organic molecules as a source of food.

The switch to predominantly aerobic metabolism is thought to have occurred in bacteria approximately 2 billion years ago, about 1.5 billion years after the first cells had formed. Aerobic eukaryotic cells (cells containing nuclei and all the other organelles) probably appeared about 1.5 billion years ago, their lineage having branched off much earlier from that of the prokaryotes. Eukaryotic cells almost certainly became aerobic by engulfing aerobic prokaryotes, with which they lived in a symbiotic relationship. The mitochondria found in both animals and plants are the descendants of such prokaryotes. Later, in branches of the eukaryotic lineage leading to plants and algae, a blue-green algaelike organism was engulfed to perform photosynthesis. It is likely that over a long period of time these organisms became the chloroplasts.

The eukaryotic cell thus apparently arose as an amalgam of different cells, in the process becoming an efficient aerobic cell whose plasma membrane was freed from energy metabolism— one of the major functions of the cell membrane of prokaryotes. The eukaryotic cell membrane was therefore able to become specialized for cell-to-cell communication and cell signaling. It may be partly for this reason that eukaryotic cells were eventually more successful at forming complex multicellular organisms than their simpler prokaryotic relatives.

*****

Cosmology

Cosmology, field of study that brings together the natural sciences, particularly astronomy and physics, in a joint effort to understand the physical universe as a unified whole.

If one looks up on a clear night, one will see that the sky is full of stars. During the summer months in the Northern Hemisphere, a faint band of light stretches from horizon to horizon, a swath of pale white cutting across a background of deepest black. For the early Egyptians, this was the heavenly Nile, flowing through the land of the dead ruled by Osiris. The ancient Greeks likened it to a river of milk. Astronomers now know that the band is actually composed of countless stars in a flattened disk seen edge on. The stars are so close to one another along the line of sight that the unaided eye has difficulty discerning the individual members. Through a large telescope, astronomers find myriads of like systems sprinkled throughout the depths of space. They call such vast collections of stars galaxies, after the Greek word for milk, and call the local galaxy to which the Sun belongs the Milky Way Galaxy or simply the Galaxy.

The Sun is a star around which Earth and the other planets revolve, and by extension every visible star in the sky is a sun in its own right. Some stars are intrinsically brighter than the Sun; others, fainter. Much less light is received from the stars than from the Sun because the stars are all much farther away. Indeed, they appear densely packed in the Milky Way only because there are so many of them. The actual separations of the stars are enormous, so large that it is conventional to measure their distances in units of how far light can travel in a given amount of time. The speed of light (in a vacuum) equals 3×10^{10} cm/sec (centimetres per second); at such a speed, it is possible to circle the Earth seven times in a single second. Thus in terrestrial terms the Sun, which lies 500 light-seconds from the Earth, is very far away; however, even the next closest star, Proxima Centauri, at a distance of 4.3 light-years (4.1×10^{18} cm), is 270,000 times farther yet. The stars that lie on the opposite side of the Milky Way from the Sun have distances that are on the order of 100,000 light-years, which is the typical diameter of a large spiral galaxy.

If the kingdom of the stars seems vast, the realm of the galaxies is larger still. The nearest galaxies to the Milky Way system are the Large and Small Magellanic Clouds, two irregular satellites of the Galaxy visible to the naked eye in the Southern Hemisphere. The Magellanic Clouds are relatively small (containing roughly 109 stars) compared to the Galaxy (with some 1011 stars), and they lie at a distance of about 200,000 light-years. The nearest large galaxy comparable to the Galaxy is the Andromeda Galaxy (also called M31 because it was the 31st entry in a catalog of astronomical objects compiled by the French astronomer Charles Messier in 1781), and it lies at a distance of about 2,000,000 light-years. The Magellanic Clouds, the Andromeda Galaxy, and the Milky Way system all are part of an aggregation of two dozen or so neighbouring galaxies known as the Local Group. The Galaxy and M31 are the largest members of this group.

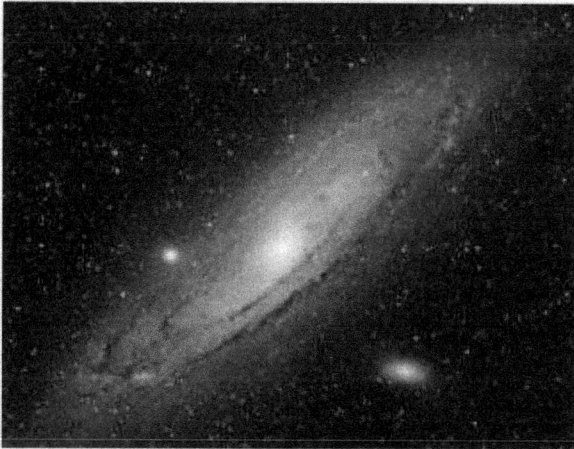

The Andromeda Galaxy, also known as the Andromeda Nebula or M31. It is the closest spiral galaxy to Earth, at a distance of 2.48 million light-years.

The Galaxy and M31 are both spiral galaxies, and they are among the brighter and more massive of all spiral galaxies. The most luminous and brightest galaxies, however, are not spirals but rather supergiant ellipticals (also called cD galaxies by astronomers for historical reasons that are not particularly illuminating). Elliptical galaxies have roundish shapes rather than the flattened distributions that characterize spiral galaxies, and they tend to occur in rich clusters (those containing thousands of members) rather than in the loose groups favoured by spirals. The brightest member galaxies of rich clusters have been detected at distances exceeding several thousand million light-years from the Earth. The branch of learning that deals with phenomena at the scale of many millions of light-years is called cosmology—a term derived from combining two Greek words, kosmos, meaning "order," "harmony," and "the world," and logos, signifying "word" or "discourse." Cosmology is, in effect, the study of the universe at large.

The Cosmological Expansion

When the universe is viewed in the large, a dramatic new feature, not present on small scales, emerges—namely, the cosmological expansion. On cosmological scales, galaxies (or, at least, clusters of galaxies) appear to be racing away from one another with the apparent velocity of recession being linearly proportional to the distance of the object. This relation is known as the Hubble law (after its discoverer, the American astronomer Edwin Powell Hubble). Interpreted in the simplest fashion, the Hubble law implies that 13.8 billion years ago all of the matter in the universe was closely packed together in an incredibly dense state and that everything then exploded in a "big bang," the signature of the explosion being written eventually in the galaxies of stars that formed out of the expanding debris of matter. Strong scientific support for this interpretation of a big bang origin of the universe comes from the detection by radio telescopes of a steady and uniform background of microwave radiation. The cosmic microwave background is believed to be a ghostly remnant of the fierce light of the primeval fireball reduced by cosmic expansion to a shadow of its former splendour but still pervading every corner of the known universe.

According to the evolutionary, or big bang, theory of the universe, the universe is expanding while the total energy and matter it contains remain constant. Therefore, as the universe expands, the density of its energy and matter must become progressively thinner. At left is a two-dimensional representation of the universe as it appears now, with galaxies occupying a typical section of space. At right, billions of years later the same amount of matter will fill a larger volume of space.

According to the evolutionary, or big bang, theory of the universe, the universe is expanding while the total energy and matter it contains remain constant. Therefore, as the universe expands, the density of its energy and matter must become progressively thinner. At left is a two-dimensional representation of the universe as it appears now, with galaxies occupying a typical section of space. At right, billions of years later the same amount of matter will fill a larger volume of space.

The simple (and most common) interpretation of the Hubble law as a recession of the galaxies over time through space, however, contains a misleading notion. In a sense, as will be made more precise later in the article, the expansion of the universe represents not so much a fundamental motion of galaxies within a framework of absolute time and absolute space, but an expansion of time and space themselves. On cosmological scales, the use of light-travel times to measure distances assumes a special significance because the lengths become so vast that even light, traveling at the fastest speed attainable by any physical entity, takes a significant fraction of the age of the universe (13.8 billion years old) to travel from an object to an observer. Thus, when astronomers measure objects at cosmological distances from the Local Group, they are seeing the objects as they existed during a time when the universe was much younger than it is today. Under these circumstances, Albert Einstein taught in his theory of general relativity that the gravitational field of everything in the universe so warps space and time as to require a very careful reevaluation of quantities whose seemingly elementary natures are normally taken for granted.

The Nature of Space and Time
Finite or infinite?

An issue that arises when one contemplates the universe at large is whether space and time are infinite or finite. After many centuries of thought by some of the best minds, humanity has still not arrived at conclusive answers to these questions. Aristotle's answer was that the material universe must be spatially finite, for if stars extended to infinity, they could not perform a complete rotation around Earth in 24 hours. Space must then itself also be finite because it is merely

136

a receptacle for material bodies. On the other hand, the heavens must be temporally infinite, without beginning or end, since they are imperishable and cannot be created or destroyed.

Except for the infinity of time, these views came to be accepted religious teachings in Europe before the period of modern science. The most notable person to publicly express doubts about restricted space was the Italian philosopher-mathematician Giordano Bruno, who asked the obvious question that, if there is a boundary or edge to space, what is on the other side? For his advocacy of an infinity of suns and earths, he was burned at the stake in 1600.

In 1610 the German astronomer Johannes Kepler provided a profound reason for believing that the number of stars in the universe had to be finite. If there were an infinity of stars, he argued, then the sky would be completely filled with them and night would not be dark! This point was rediscussed by the astronomers Edmond Halley of England and Jean-Philippe-Loys de Chéseaux of Switzerland in the 18th century, but it was not popularized as a paradox until Wilhelm Olbers of Germany took up the problem in the 19th century. The difficulty became potentially very real with American astronomer Edwin Hubble's measurement of the enormous extent of the universe of galaxies with its large-scale homogeneity and isotropy. His discovery of the systematic recession of the galaxies provided an escape, however. At first people thought that the redshift effect alone would suffice to explain why the sky is dark at night—namely, that the light from the stars in distant galaxies would be redshifted to long wavelengths beyond the visible regime. The modern consensus is, however, that a finite age for the universe is a far more important effect. Even if the universe is spatially infinite, photons from very distant galaxies simply do not have the time to travel to Earth because of the finite speed of light. There is a spherical surface, the cosmic event horizon (13.8 billion light-years in radial distance from Earth at the current epoch), beyond which nothing can be seen even in principle; and the number (roughly 10^{10}) of galaxies within this cosmic horizon, the observable universe, are too few to make the night sky bright.

When one looks to great distances, one is seeing things as they were a long time ago, again because light takes a finite time to travel to Earth. Over such great spans, do the classical notions of Euclid concerning the properties of space necessarily continue to hold? The answer given by Einstein was: No, the gravitation of the mass contained in cosmologically large regions may warp one's usual perceptions of space and time; in particular, the Euclidean postulate that parallel lines never cross need not be a correct description of the geometry of the actual universe. And in 1917 Einstein presented a mathematical model of the universe in which the total volume of space was finite yet had no boundary or edge. The model was based on his theory of general relativity that utilized a more generalized approach to geometry devised in the 19th century by the German mathematician Bernhard Riemann.

Gravitation And The Geometry of Space-Time

The physical foundation of Einstein's view of gravitation, general relativity, lies on two empirical findings that he elevated to the status of basic postulates. The first postulate is the relativity principle: local physics is governed by the theory of special relativity. The second postulate is the equivalence principle: there is no way for an observer to distinguish locally between gravity and acceleration. The motivation for the second postulate comes from Galileo's observation that all objects—independent of mass, shape, colour, or any other property—accelerate at the same rate in a (uniform) gravitational field.

Einstein's theory of special relativity, which he developed in 1905, had as its basic premises (1) the notion (also dating back to Galileo) that the laws of physics are the same for all inertial observers and (2) the constancy of the speed of light in a vacuum—namely, that the speed of light has the same value (3 $\times$ 10^{10} centimetres per second [cm/sec], or 2 $\times$ 10^5 miles per second [miles/sec]) for all inertial observers independent of their motion relative to the source of the light. Clearly, this second premise is incompatible with Euclidean and Newtonian precepts of absolute space and absolute time, resulting in a program that merged space and time into a single structure, with well-known consequences. The space-time structure of special relativity is often called "flat" because, among other things, the propagation of photons is easily represented on a flat sheet of graph paper with equal-sized squares. Let each tick on the vertical axis represent one light-year (9.46 $\times$ 10^{17} cm [5.88 $\times$ 10^{12} miles]) of distance in the direction of the flight of the photon, and each tick on the horizontal axis represent the passage of one year (3.16 $\times$ 10^7 seconds) of time. The propagation path of the photon is then a 45° line because it flies one light-year in one year (with respect to the space and time measurements of all inertial observers no matter how fast they move relative to the photon).

The principle of equivalence in general relativity allows the locally flat space-time structure of special relativity to be warped by gravitation, so that (in the cosmological case) the propagation of the photon over thousands of millions of light-years can no longer be plotted on a globally flat sheet of paper. To be sure, the curvature of the paper may not be apparent when only a small piece is examined, thereby giving the local impression that space-time is flat (i.e., satisfies special relativity). It is only when the graph paper is examined globally that one realizes it is curved (i.e., satisfies general relativity).

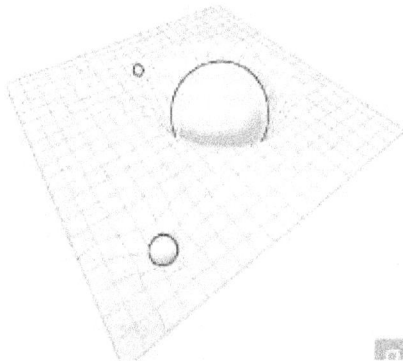

Curved space-timeThe four dimensional space-time continuum itself is distorted in the vicinity of any mass, with the amount of distortion depending on the mass and the distance from the mass. Thus, relativity accounts for Newton's inverse square law of gravity through geometry and thereby does away with the need for any mysterious "action at a distance."

In Einstein's 1917 model of the universe, the curvature occurs only in space, with the graph paper being rolled up into a cylinder on its side, a loop around the cylinder at constant time having a circumference of 2πR—the total spatial extent of the universe. Notice that the "radius of the universe" is measured in a "direction" perpendicular to the space-time surface of the graph paper. Since the ringed space axis corresponds to one of three dimensions of the actual world (any will do since all directions are equivalent in an isotropic model), the radius of the universe exists in a fourth spatial dimension (not time) which is not part of the real world. This fourth spatial dimension is a mathematical artifice introduced to represent diagrammatically the solution (in this case) of equations for curved three-dimensional space that need not refer to any dimensions other than the three physical ones. Photons traveling in a straight line in any physical direction have trajectories that go diagonally (at 45° angles to the space and time axes) from corner to corner of each little square cell of the space-time grid; thus, they describe helical paths on the cylindrical surface of the graph paper, making one turn after traveling a spatial distance $2\pi R$. In other words, always flying dead ahead, photons would return to where they started from after going a finite distance without ever coming to an edge or boundary. The distance to the "other side" of the universe is therefore πR, and it would lie in any and every direction; space would be closed on itself.

Now, except by analogy with the closed two-dimensional surface of a sphere that is uniformly curved toward a centre in a third dimension lying nowhere on the two-dimensional surface, no three-dimensional creature can visualize a closed three-dimensional volume that is uniformly curved toward a centre in a fourth dimension lying nowhere in the three-dimensional volume. Nevertheless, three-dimensional creatures could discover the curvature of their three-dimensional world by performing surveying experiments of sufficient spatial scope. They could draw circles, for example, by tacking down one end of a string and tracing along a single plane the locus described by the other end when the string is always kept taut in between (a straight line) and walked around by a surveyor. In Einstein's universe, if the string were short compared to the quantity R, the circumference of the circle divided by the length of the string (the circle's radius) would nearly equal $2\pi = 6.2837853...$, thereby fooling the three-dimensional creatures into thinking that Euclidean geometry gives a correct description of their world. However, the ratio of circumference to length of string would become less than 2π when the length of string became comparable to R. Indeed, if a string of length πR could be pulled taut to the antipode of a positively curved universe, the ratio would go to zero. In short, at the tacked-down end the string could be seen to sweep out a great arc in the sky from horizon to horizon and back again; yet, to make

140

the string do this, the surveyor at the other end need only walk around a circle of vanishingly small circumference.

To understand why gravitation can curve space (or more generally, space-time) in such startling ways, consider the following thought experiment that was originally conceived by Einstein. Imagine an elevator in free space accelerating upward, from the viewpoint of a woman in inertial space, at a rate numerically equal to g, the gravitational field at the surface of Earth. Let this elevator have parallel windows on two sides, and let the woman shine a brief pulse of light toward the windows. She will see the photons enter close to the top of the near window and exit near the bottom of the far window because the elevator has accelerated upward in the interval it takes light to travel across the elevator. For her, photons travel in a straight line, and it is merely the acceleration of the elevator that has caused the windows and floor of the elevator to curve up to the flight path of the photons.

Let there now be a man standing inside the elevator. Because the floor of the elevator accelerates him upward at a rate g, he may—if he chooses to regard himself as stationary—think that he is standing still on the surface of Earth and is being pulled to the ground by its gravitational field g. Indeed, in accordance with the equivalence principle, without looking out the windows (the outside is not part of his local environment), he cannot perform any local experiment that would inform him otherwise. Let the woman shine her pulse of light. The man sees, just like the woman, that the photons enter near the top edge of one window and exit near the bottom of the other. And just like the woman, he knows that photons propagate in straight lines in free space. (By the relativity principle, they must agree on the laws of physics if they are both inertial observers.) However, since he actually sees the photons follow a curved path relative to himself, he concludes that they must be bent by the force of gravity. The woman tries to tell him there is no such force at work; he is not an inertial observer. Nonetheless, he has the solidity of Earth beneath him, so he insists on attributing his acceleration to the force of gravity. According to Einstein, they are both right. There is no need to distinguish locally between acceleration and gravity—the two are in some sense equivalent. But if that is the case, then it must be true that gravity—"real" gravity—can actually bend light. And indeed it can, as many experiments have shown since Einstein's first discussion of the phenomenon.

It was the genius of Einstein to go even further. Rather than speak of the force of gravitation having bent the photons into a curved path, might it not be more fruitful to think of photons as always flying in straight lines—in the sense that a straight line is the shortest distance between two points—and that what really happens is that gravitation bends space-time? In other words, perhaps gravitation is curved space-time, and photons fly along the shortest paths possible in this curved space-time, thus giving the appearance of being bent by a "force" when one insists on thinking that space-time is flat. The utility of taking this approach is that it becomes automatic that all test bodies fall at the same rate under the "force" of gravitation, for they are merely producing their natural trajectories in a background space-time that is curved in a certain fashion independent of the test bodies. What was a minor miracle for Galileo and Newton becomes the most natural thing in the world for Einstein.

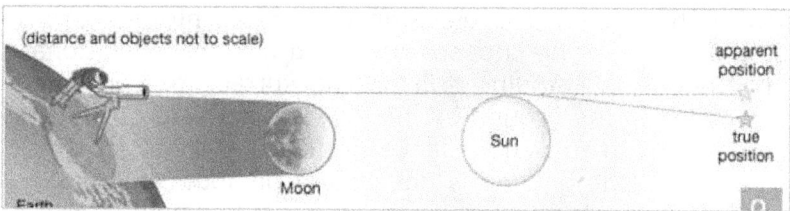

(distance and objects not to scale)

Experimental evidence for general relativityIn 1919, observation of a solar eclipse confirmed Einstein's prediction that light is bent in the presence of mass. This experimental support for his general theory of relativity garnered him instant worldwide acclaim.

To complete the program and to conform with Newton's theory of gravitation in the limit of weak curvature (weak field), the source of space-time curvature would have to be ascribed to mass (and energy). The mathematical expression of these ideas constitutes Einstein's theory of general relativity, one of the most beautiful artifacts of pure thought ever produced. The American physicist John Archibald Wheeler and his colleagues summarized Einstein's view of the universe in these terms:

- Curved spacetime tells mass-energy how to move;
- mass-energy tells spacetime how to curve.

Contrast this with Newton's view of the mechanics of the heavens:

- Force tells mass how to accelerate;
- mass tells gravity how to exert force.

Notice therefore that Einstein's worldview is not merely a quantitative modification of Newton's picture (which is also possible via an equivalent route using the methods of quantum field theory) but represents a qualitative change of perspective. And modern experiments have amply justified the fruitfulness of Einstein's alternative interpretation of gravitation as geometry rather than as force. His theory would have undoubtedly delighted the Greeks.

Relativistic Cosmologies
Einstein's model

To derive his 1917 cosmological model, Einstein made three assumptions that lay outside the scope of his equations. The first was to suppose that the universe is homogeneous and isotropic in the large (i.e., the same everywhere on average at any instant in time), an assumption that the English astrophysicist Edward A. Milne later elevated to an entire philosophical outlook by naming it the cosmological principle. Given the success of the Copernican revolution, this outlook is a natural one. Newton himself had it implicitly in mind when he took the initial state of the universe to be everywhere the same before it developed "ye Sun and Fixt stars."

The second assumption was to suppose that this homogeneous and isotropic universe had a closed spatial geometry. As described above, the total volume of a three-dimensional space with uniform positive curvature would be finite but possess no edges or boundaries (to be consistent with the first assumption).

The third assumption made by Einstein was that the universe as a whole is static—i.e., its large-scale properties do not vary with time. This assumption, made before Hubble's observational discovery of the expansion of the universe, was also natural; it was the simplest approach, as Aristotle had discovered, if one wishes to avoid a discussion of a creation event. Indeed, the philosophical attraction of the notion that the universe on average is not only homogeneous and isotropic in space but also constant in time was so appealing that a school of English cosmologists—Hermann Bondi, Fred Hoyle, and Thomas Gold—would call it the perfect cosmological principle and carry its implications in the 1950s to the ultimate refinement in the so-called steady-state theory.

To his great chagrin Einstein found in 1917 that with his three adopted assumptions, his equations of general relativity—as originally written down—had no meaningful solutions. To obtain a solution, Einstein realized that he had to add to his equations an extra term, which came to be called the cosmological constant. If one speaks in Newtonian terms, the cosmological constant could be interpreted as a repulsive force of unknown origin that could exactly balance the attraction of gravitation of all the matter in Einstein's closed universe and keep it from moving. The inclusion of such a term in a more general context, however, meant that the universe in the absence of any mass-energy (i.e., consisting of a vacuum) would not have a space-time structure that was flat (i.e., would not have satisfied the dictates of special relativity exactly). Einstein was prepared to make such a sacrifice only very reluctantly, and, when he later learned of Hubble's discovery of the expansion of the universe and realized that he could have predicted it had he only had more faith in the original form of his equations, he regretted the introduction of the cosmological constant as the "biggest blunder" of his life. Ironically, observations of distant supernovas have shown the existence of dark energy, a repulsive force that is the dominant component of the universe.

De Sitter's Model

It was also in 1917 that the Dutch astronomer Willem de Sitter recognized that he could obtain a static cosmological model differing from Einstein's simply by removing all matter. The solution remains stationary essentially because there is no matter to move about. If some test particles are reintroduced into the model, the cosmological term would propel them away from each other. Astronomers now began to wonder if this effect might not underlie the recession of the spiral galaxies.

Friedmann-Lemaître Models

In 1922 Aleksandr A. Friedmann, a Russian meteorologist and mathematician, and in 1927 Georges Lemaître, a Belgian cleric, independently discovered solutions to Einstein's equations that contained realistic amounts of matter. These evolutionary models correspond to big bang cosmologies. Friedmann and Lemaître adopted Einstein's assumption of spatial homogeneity and isotropy (the cosmological principle). They rejected, however, his assumption of time independence and considered both positively curved spaces ("closed" universes) as well as negatively curved spaces ("open" universes). The difference between the approaches of Friedmann and Lemaître is that the former set the cosmological constant equal to zero, whereas the latter retained the possibility that it might have a nonzero value. To simplify the discussion, only the Friedmann models are considered here.

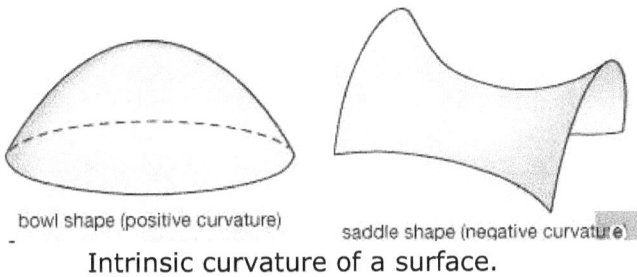

bowl shape (positive curvature) saddle shape (negative curvature)

Intrinsic curvature of a surface.

The decision to abandon a static model meant that the Friedmann models evolve with time. As such, neighbouring pieces of matter have recessional (or contractional) phases when they separate from (or approach) one another with an apparent velocity that increases linearly with increasing distance. Friedmann's models thus anticipated Hubble's law before it had been formulated on an observational basis. It was Lemaître, however, who had the good fortune of deriving the results at the time when the recession of the galaxies was being recognized as a fundamental cosmological observation, and it was he who clarified the theoretical basis for the phenomenon.

The geometry of space in Friedmann's closed models is similar to that of Einstein's original model; however, there is a curvature to time as well as one to space. Unlike Einstein's model, where time runs eternally at each spatial point on an uninterrupted horizontal line that extends infinitely into the past and future, there is a beginning and end to time in Friedmann's version of a closed universe when material expands from or is recompressed to infinite densities. These instants are called the instants of the "big bang" and the "big squeeze," respectively.

The global space-time diagram for the middle half of the expansion-compression phases can be depicted as a barrel lying on its side. The space axis corresponds again to any one direction in the universe, and it wraps around the barrel. Through each spatial point runs a time axis that extends along the length of the barrel on its (space-time) surface. Because the barrel is curved in both space and time, the little squares in the grid of the curved sheet of graph paper marking the space-time surface are of nonuniform size, stretching to become bigger when the barrel broadens (universe expands) and shrinking to become smaller when the barrel narrows (universe contracts).

It should be remembered that only the surface of the barrel has physical significance; the dimension off the surface toward the axle of the barrel represents the fourth spatial dimension, which is not part of the real three-dimensional world. The space axis circles the barrel and closes upon itself after traversing a circumference equal to $2\pi R$, where R, the radius of the universe (in the fourth dimension), is now a function of the time t. In a closed Friedmann model, R starts equal to zero at time t = 0 (not shown in barrel diagram), expands to a maximum value at time t = tm (the middle of the barrel), and recontracts to zero (not shown) at time t = 2tm, with the value of tm dependent on the total amount of mass that exists in the universe.

Imagine now that galaxies reside on equally spaced tick marks along the space axis. Each galaxy on average does not move spatially with respect to its tick mark in the spatial (ringed) direction but is carried forward horizontally by the march of time. The total number of galaxies on the spatial ring is conserved as time changes, and therefore their average spacing increases or decreases as the total circumference $2\pi R$ on the ring increases or decreases (during the expansion or contraction phases). Thus, without in a sense actually moving in the spatial direction, galaxies can be carried apart by the expansion of space itself. From this point of view, the recession of galaxies is not a "velocity" in the usual sense of the word. For example, in a closed Friedmann model, there could be galaxies that started, when R was small, very close to the Milky Way system on the opposite side of the universe. Now, 1010 years later, they are still on the opposite side of the universe but at a distance much greater than 1010 light-years away. They reached those distances without ever having had to move (relative to any local observer) at speeds faster than light—

indeed, in a sense without having had to move at all. The separation rate of nearby galaxies can be thought of as a velocity without confusion in the sense of Hubble's law, if one wants, but only if the inferred velocity is much less than the speed of light.

On the other hand, if the recession of the galaxies is not viewed in terms of a velocity, then the cosmological redshift cannot be viewed as a Doppler shift. How, then, does it arise? The answer is contained in the barrel diagram when one notices that, as the universe expands, each small cell in the space-time grid also expands. Consider the propagation of electromagnetic radiation whose wavelength initially spans exactly one cell length (for simplicity of discussion), so that its head lies at a vertex and its tail at one vertex back. Suppose an elliptical galaxy emits such a wave at some time $t1$. The head of the wave propagates from corner to corner on the little square grids that look locally flat, and the tail propagates from corner to corner one vertex back. At a later time $t2$, a spiral galaxy begins to intercept the head of the wave. At time $t2$, the tail is still one vertex back, and therefore the wave train, still containing one wavelength, now spans one current spatial grid spacing. In other words, the wavelength has grown in direct proportion to the linear expansion factor of the universe. Since the same conclusion would have held if n wavelengths had been involved instead of one, all electromagnetic radiation from a given object will show the same cosmological redshift if the universe (or, equivalently, the average spacing between galaxies) was smaller at the epoch of transmission than at the epoch of reception. Each wavelength will have been stretched in direct proportion to the expansion of the universe in between.

A nonzero peculiar velocity for an emitting galaxy with respect to its local cosmological frame can be taken into account by Doppler-shifting the emitted photons before applying the cosmological redshift factor; i.e., the observed redshift would be a product of two factors. When the observed redshift is large, one usually assumes that the dominant contribution is of cosmological origin. When this assumption is valid, the redshift is a monotonic function of both distance and time during the expansional phase of any cosmological model. Thus, astronomers often use the redshift z as a shorthand indicator of both distance and elapsed time. Following from this, the statement "object X lies at $z = a$" means that "object X lies at a distance associated with redshift a"; the statement "event Y occurred at redshift z = b" means that "event Y occurred a time ago associated with redshift b."

The open Friedmann models differ from the closed models in both spatial and temporal behaviour. In an open universe the total volume of space and the number of galaxies contained in it are infinite. The three-dimensional spatial geometry is one of uniform negative curvature in the sense that, if circles are drawn with very large lengths of string, the ratio of circumferences to lengths of string are greater than 2π. The temporal history begins again with expansion from a big bang of infinite density, but now the expansion continues indefinitely, and the average density of matter and radiation in the universe would eventually become vanishingly small. Time in such a model has a beginning but no end.

The Einstein–de Sitter Universe

In 1932 Einstein and de Sitter proposed that the cosmological constant should be set equal to zero, and they derived a homogeneous and isotropic model that provides the separating case between the closed and open Friedmann models; i.e., Einstein and de Sitter assumed that the spatial curvature of the universe is neither positive nor negative but rather zero. The spatial geometry of the Einstein–de Sitter universe is Euclidean (infinite total volume), but space-time is not globally flat (i.e., not exactly the space-time of special relativity). Time again commences with a big bang and the galaxies recede forever, but the recession rate (Hubble's "constant") asymptotically coasts to zero as time advances to infinity. Because the geometry of space and the gross evolutionary properties are uniquely defined in the Einstein–de Sitter model, many people with a philosophical bent long considered it the most fitting candidate to describe the actual universe.

Bound And Unbound Universes And The Closure Density

The different separation behaviours of galaxies at large timescales in the Friedmann closed and open models and the Einstein–de Sitter model allow a different classification scheme than one based on the global structure of space-time. The alternative way of looking at things is in terms of gravitationally bound and unbound systems: closed models where galaxies initially separate but later come back together again represent bound universes; open models where galaxies continue to separate forever represent unbound universes; the Einstein–de Sitter model where galaxies separate forever but slow to a halt at infinite time represents the critical case.

How the relative size of the universe changes with time in four different models. The red line shows a universe devoid of matter, with constant expansion. Pink shows a collapsing universe, with six times the critical density of matter. Green shows a model favoured until 1998, with exactly the critical density and a universe 100 percent matter. Blue shows the currently favoured scenario, with exactly the critical density, of which 27 percent is visible and dark matter and 73 percent is dark energy.

How the relative size of the universe changes with time in four different models. The red line shows a universe devoid of matter, with constant expansion. Pink shows a collapsing universe, with six times the critical density of matter. Green shows a model favoured until 1998, with exactly the critical density and a universe 100 percent matter. Blue shows the currently favoured scenario, with exactly the critical density, of which 27 percent is visible and dark matter and 73 percent is dark energy.

The advantage of this alternative view is that it focuses attention on local quantities where it is possible to think in the simpler terms of Newtonian physics—attractive forces, for example. In this picture it is intuitively clear that the feature that should distinguish whether or not gravity is capable of bringing a given expansion rate to a halt depends on the amount of mass (per unit volume) present. This is indeed the case; the Newtonian and relativistic formalisms give the same criterion for the critical, or closure, density (in mass equivalent of matter and radiation) that separates closed or bound universes from open or unbound ones. If Hubble's constant at the present epoch is denoted as $H0$, then the closure density (corresponding to an Einstein–de Sitter model) equals $3H02/8\pi G$, where G is the universal gravitational constant in both Newton's and Einstein's theories of gravity. The numerical value of Hubble's constant $H0$ is 22 kilometres per second per million light-years; the closure density then equals $10-29$ gram per cubic centimetre, the equivalent of about six hydrogen atoms on average per cubic metre of cosmic space. If the actual cosmic average is greater than this value, the universe is bound (closed) and, though currently expanding, will end in a crush of unimaginable proportion. If it is less, the universe is unbound (open) and will expand forever. The result is intuitively plausible since the smaller the mass density, the smaller the role for gravitation, so the more the universe will approach free expansion (assuming that the cosmological constant is zero).

The mass in galaxies observed directly, when averaged over cosmological distances, is estimated to be only a few percent of the amount required to close the universe. The amount contained in the radiation field (most of which is in the cosmic microwave background) contributes negligibly to the total at present. If this were all, the universe would be open and unbound. However, the dark matter that has been deduced from various dynamic arguments is about 23 percent of the universe, and dark energy supplies the remaining amount, bringing the total average mass density up to 100 percent of the closure density.

The Hot Big Bang

Given the measured radiation temperature of 2.735 kelvins (K), the energy density of the cosmic microwave background can be shown to be about 1,000 times smaller than the average rest-energy density of ordinary matter in the universe. Thus, the current universe is matter-dominated. If one goes back in time to redshift z, the average number densities of particles and photons were both bigger by the same factor $(1 + z)^3$ because the universe was more compressed by this factor, and the ratio of these two numbers would have maintained its current value of about one hydrogen nucleus, or proton, for every 109 photons. The wavelength of each photon, however, was shorter by the factor $1 + z$ in the past than it is now; therefore, the energy density of radiation increases faster by one factor of $1 + z$ than the rest-energy density of matter. Thus, the radiation energy density becomes comparable to the energy density of ordinary matter at a redshift of about 1,000. At redshifts larger than 10,000, radiation would have dominated even over the dark matter of the universe. Between these two values radiation would have decoupled from matter when hydrogen recombined. It is not possible to use photons to observe redshifts larger than about 1,090, because the cosmic plasma at temperatures above 4,000 K is essentially opaque before recombination. One can think of the spherical surface as an inverted "photosphere" of the observable universe. This spherical surface of last scattering probably has slight ripples in it that account for the slight anisotropies observed in the cosmic microwave background today. In any case, the earliest stages of the universe's history—for example, when temperatures were 109 K and higher—cannot be examined by light received through any telescope. Clues must be sought by comparing the matter content with theoretical calculations.

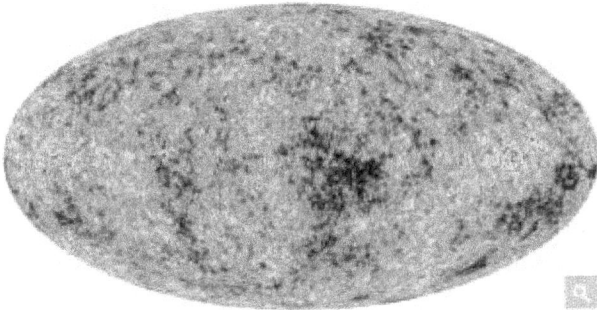

Wilkinson Microwave Anisotropy Probe

A full-sky map produced by the Wilkinson Microwave Anisotropy Probe (WMAP) showing cosmic background radiation, a very uniform glow of microwaves emitted by the infant universe more than 13 billion years ago. Colour differences indicate tiny fluctuations in the intensity of the radiation, a result of tiny variations in the density of matter in the early universe. According to inflation theory, these irregularities were the "seeds" that became the galaxies. WMAP's data support the big bang and inflation models.

For this purpose, fortunately, the cosmological evolution of model universes is especially simple and amenable to computation at redshifts much larger than 10,000 (or temperatures substantially above 30,000 K) because the physical properties of the dominant component, photons, then are completely known. In a radiation-dominated early universe, for example, the radiation temperature T is very precisely known as a function of the age of the universe, the time t after the big bang.

Primordial Nucleosynthesis
According to the considerations outlined above, at a time t less than 10-4 seconds, the creation of matter-antimatter pairs would have been in thermodynamic equilibrium with the

ambient radiation field at a temperature T of about 1012 K. Nevertheless, there was a slight excess of matter particles (e.g., protons) compared to antimatter particles (e.g., antiprotons) of roughly a few parts in 109. This is known because, as the universe aged and expanded, the radiation temperature would have dropped and each antiproton and each antineutron would have annihilated with a proton and a neutron to yield two gamma rays; and later each antielectron would have done the same with an electron to give two more gamma rays. After annihilation, however, the ratio of the number of remaining protons to photons would be conserved in the subsequent expansion to the present day. Since that ratio is known to be one part in 109, it is easy to work out that the original matter-antimatter asymmetry must have been a few parts per 10^9.

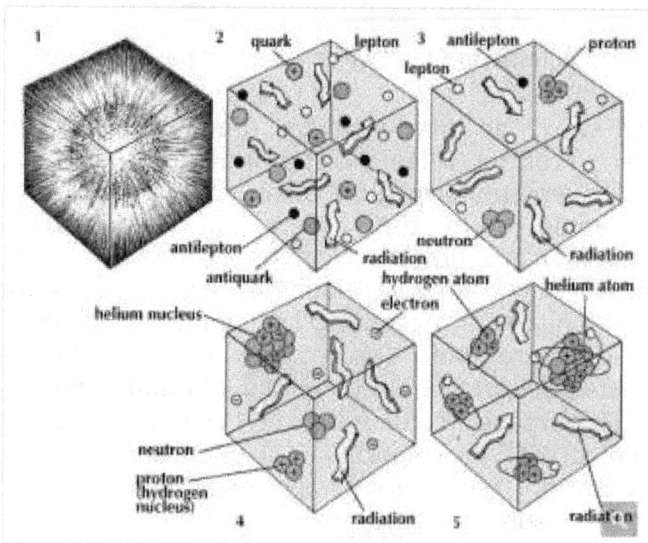

Immediately after the big bang (1), the universe was filled with a dense "soup" of subatomic particles (2), called quarks and leptons (such as electrons), and their antiparticle equivalents. By 0.01 second after the big bang (3), some of the quarks had united to form neutrons and protons. (After another 2 seconds, the only leptons remaining were electrons; the antiparticles had been annihilated.) After 3.5 minutes (4), hydrogen and helium nuclei had formed. After a million years (5), the universe was populated with hydrogen and helium atoms, the raw material of stars and galaxies. The initial radiation from the big bang had grown less energetic.

In any case, after proton-antiproton and neutron-antineutron annihilation but before electron-antielectron annihilation, it is possible to calculate that for every excess neutron there were about five excess protons in thermodynamic equilibrium with one another through neutrino and antineutrino interactions at a temperature of about 1010 K. When the universe reached an age of a few seconds, the temperature would have dropped significantly below 1010 K, and electron-antielectron annihilation would have occurred, liberating the neutrinos and antineutrinos to stream freely through the universe. With no neutrino-antineutrino reactions to replenish their supply, the neutrons would have started to decay with a half-life of 10.6 minutes to protons and electrons (and antineutrinos). However, at an age of 1.5 minutes, well before neutron decay went to completion, the temperature would have dropped to 109 K, low enough to allow neutrons to be captured by protons to form a nucleus of heavy hydrogen, or deuterium. (Before that time, the reaction could still have taken place, but the deuterium nucleus would immediately have broken up under the prevailing high temperatures.) Once deuterium had formed, a very fast chain of reactions set in, quickly assembling most of the neutrons and deuterium nuclei with protons to yield helium nuclei. If the decay of neutrons is ignored, an original mix of 10 protons and two neutrons (one neutron for every five protons) would have assembled into one helium nucleus (two protons plus two neutrons), leaving more than eight protons (eight hydrogen nuclei). This amounts to a helium-mass fraction of $4/12 = 1/3$—i.e., 33 percent. A more sophisticated calculation that takes into account the concurrent decay of neutrons and other complications yields a helium-mass fraction in the neighbourhood of 25 percent and a hydrogen-mass fraction of 75 percent, which are close to the deduced primordial values from astronomical observations. This agreement provides one of the primary successes of hot big bang theory.

The Deuterium Abundance

Not all of the deuterium formed by the capture of neutrons by protons would be further reacted to produce helium. A small residual can be expected to remain, the exact fraction depending sensitively on the density of ordinary matter existing in the universe when the universe was a few minutes old. The problem can be turned around: given measured values of the deuterium abundance (corrected for various effects), what density of ordinary matter needs to be present at a temperature of 109 K so that the nuclear reaction calculations will reproduce the measured deuterium abundance? The answer is known, and this density of ordinary matter can be expanded by simple scaling relations from a radiation temperature of 109 K to one of 2.735 K. This yields a predicted present density of ordinary matter and can be compared with the density inferred to exist in galaxies when averaged over large regions. The two numbers are within a factor of a few of each other. In other words, the deuterium calculation implies much of the ordinary matter in the universe has already been seen in observable galaxies. Ordinary matter cannot be the hidden mass of the universe

The Very Early Universe
Inhomogeneous Nucleosynthesis

One possible modification concerns models of so-called inhomogeneous nucleosynthesis. The idea is that in the very early universe (the first microsecond) the subnuclear particles that later made up the protons and neutrons existed in a free state as a quark-gluon plasma. As the universe expanded and cooled, this quark-gluon plasma would undergo a phase transition and become confined to protons and neutrons (three quarks each). In laboratory experiments of similar phase transitions—for example, the solidification of a liquid into a solid—involving two or more substances, the final state may contain a very uneven distribution of the constituent substances, a fact exploited by industry to purify certain materials. Some astrophysicists have proposed that a similar partial separation of neutrons and protons may have occurred in the very early universe. Local pockets where protons abounded may have few neutrons and vice versa for where neutrons abounded. Nuclear reactions may then have occurred much less efficiently per proton and neutron nucleus than accounted for by standard calculations, and the average density of matter may be correspondingly increased—perhaps even to the point where ordinary matter can close the present-day universe. Unfortunately, calculations carried out under the inhomogeneous hypothesis seem to indicate that conditions leading to the correct proportions of deuterium and helium-4 produce too much primordial lithium-7 to be compatible with measurements of the atmospheric compositions of the oldest stars.

Matter-Antimatter Asymmetry

A curious number that appeared in the above discussion was the few parts in 109 asymmetry initially between matter and antimatter (or equivalently, the ratio 10^{-9} of protons to photons in the present universe). What is the origin of such a number—so close to zero yet not exactly zero?

At one time the question posed above would have been considered beyond the ken of physics, because the net "baryon" number (for present purposes, protons and neutrons minus antiprotons and antineutrons) was thought to be a conserved quantity. Therefore, once it exists, it always exists, into the indefinite past and future. Developments in particle physics during the 1970s, however, suggested that the net baryon number may in fact undergo alteration. It is certainly very nearly maintained at the relatively low energies accessible in terrestrial experiments, but it may not be conserved at the almost arbitrarily high energies with which particles may have been endowed in the very early universe.

An analogy can be made with the chemical elements. In the 19th century most chemists believed the elements to be strictly conserved quantities; although oxygen and hydrogen atoms can be combined to form water molecules, the original oxygen and hydrogen atoms can always be recovered by chemical or physical means. However, in the 20th century with the discovery and elucidation of nuclear forces, chemists came to realize that the elements are conserved if they are subjected only to chemical forces (basically electromagnetic in origin); they can be transmuted by the introduction of nuclear forces, which enter characteristically only when much higher energies per particle are available than in chemical reactions.

In a similar manner it turns out that at very high energies new forces of nature may enter to transmute the net baryon number. One hint that such a transmutation may be possible lies in the remarkable fact that a proton and an electron seem at first sight to be completely different entities, yet they have, as far as one can tell to very high experimental precision, exactly equal but opposite electric charges. Is this a fantastic coincidence, or does it represent a deep physical connection? A connection would obviously exist if it can be shown, for example, that a proton is capable of decaying into a positron (an antielectron) plus electrically neutral particles. Should this be possible, the proton would necessarily have the same charge as the positron, for charge is exactly conserved in all reactions. In turn, the positron would necessarily have the opposite charge of the electron, as it is its antiparticle. Indeed, in some sense the proton (a baryon) can even be said to be merely the "excited" version of an antielectron (an "antilepton").

Motivated by this line of reasoning, experimental physicists searched hard during the 1980s for evidence of proton decay. They found none and set a lower limit of 1032 years for the lifetime of the proton if it is unstable. This value is greater than what theoretical physicists had originally predicted on the basis of early unification schemes for the forces of nature. Later versions can accommodate the data and still allow the proton to be unstable. Despite the inconclusiveness of the proton-decay experiments, some of the apparatuses were eventually put to good astronomical use. They were converted to neutrino detectors and provided valuable information on the solar neutrino problem, as well as giving the first positive recordings of neutrinos from a supernova explosion (namely, supernova 1987A).

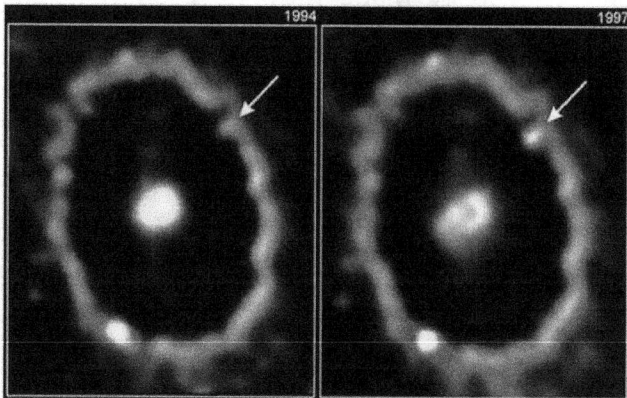

Supernova 1987A in the Large Magellanic Cloud. This picture shows the faint outer rings and bright inner ring characteristic of an hourglass nebula.

With respect to the cosmological problem of the matter-antimatter asymmetry, one theoretical approach is founded on the idea of a grand unified theory (GUT), which seeks to explain the electromagnetic, weak nuclear, and strong nuclear forces as a single grand force of nature. This approach suggests that an initial collection of very heavy particles, with zero baryon and lepton number, may decay into many lighter particles (baryons and leptons) with the desired average for the net baryon number (and net lepton number) of a few parts per 109. This event is supposed to have occurred at a time when the universe was perhaps 10–35 second old.

Another approach to explaining the asymmetry relies on the process of CP violation, or violation of the combined conservation laws associated with charge conjugation (C) and parity (P) by the weak force, which is responsible for reactions such as the radioactive decay of atomic nuclei. Charge conjugation implies that every charged particle has an oppositely charged antimatter counterpart, or antiparticle. Parity conservation means that left and right and up and down are indistinguishable in the sense that an atomic nucleus emits decay products up as often as down and left as often as right. With a series of debatable but plausible assumptions, it can be demonstrated that the observed imbalance or asymmetry in the matter-antimatter ratio may have been produced by the occurrence of CP violation in the first seconds after the big bang. CP violation is expected to be more prominent in the decay of particles known as B-mesons. In 2010, scientists at the Fermi National Acclerator Laboratory in Batavia, Ill., finally detected a slight preference for B-mesons to decay into muons rather than anti-muons.

Superunification And The Planck Era

Why should a net baryon fraction initially of zero be more appealing aesthetically than 10–9? The underlying motivation here is perhaps the most ambitious undertaking ever attempted in the history of science—the attempt to explain the creation of truly everything from literally nothing. In other words, is the creation of the entire universe from a vacuum possible?

The evidence for such an event lies in another remarkable fact. It can be estimated that the total number of protons in the observable universe is an integer 80 digits long. No one of course knows all 80 digits, but for the argument about to be presented, it suffices only to know that they exist. The total number of electrons in the observable universe is also an integer 80 digits long. In all likelihood these two integers are equal, digit by digit—if not exactly, then very nearly so. This inference comes from the fact that, as far as astronomers can tell, the total electric charge in the universe is zero (otherwise electrostatic forces would overwhelm gravitational forces). Is this another coincidence, or does it represent a deeper connection? The apparent coincidence becomes trivial if the entire universe was created from a vacuum since a vacuum has by definition zero electric charge. It is a truism that one cannot get something for nothing. The interesting question is whether one can get everything for nothing. Clearly, this is a very speculative topic for scientific investigation, and the ultimate answer depends on a sophisticated interpretation of what "nothing" means.

The words "nothing," "void," and "vacuum" usually suggest uninteresting empty space. To modern quantum physicists, however, the vacuum has turned out to be rich with complex and unexpected behaviour. They envisage it as a state of minimum energy where quantum fluctuations, consistent with the uncertainty principle of the German physicist Werner Heisenberg, can lead to the temporary formation of particle-antiparticle pairs. In flat space-time, destruction follows closely upon creation (the pairs are said to be virtual) because there is no source of energy to give the pair permanent existence. All the known forces of nature acting between a particle and antiparticle are attractive and will pull the pair together to annihilate one another. In the expanding space-time of the very early universe, however, particles and antiparticles may separate and become part of the observable world. In other words, sharply curved space-time can give rise to the creation of real pairs with positive mass-energy, a fact first demonstrated in the context of black holes by the English astrophysicist Stephen W. Hawking.

Yet Einstein's picture of gravitation is that the curvature of space-time itself is a consequence of mass-energy. Now, if curved space-time is needed to give birth to mass-energy and if mass-energy is needed to give birth to curved space-time, which came first, space-time or mass-energy? The suggestion that they both rose from something still more fundamental raises a new question: What is more fundamental than space-time and mass-energy? What can give rise to both mass-energy and space-time? No one knows the answer to this question, and perhaps some would argue that the answer is not to be sought within the boundaries of natural science.

Hawking and the American cosmologist James B. Hartle have proposed that it may be possible to avert a beginning to time by making it go imaginary (in the sense of the mathematics of complex numbers) instead of letting it suddenly appear or disappear. Beyond a certain point in their scheme, time may acquire the characteristic of another spatial dimension rather than refer to some sort of inner clock. Another proposal states that, when space and time approach small enough values (the Planck values; see below), quantum effects make it meaningless to ascribe any classical notions to their properties. The most promising approach to describe the situation comes from the theory of "superstrings."

Superstrings represent one example of a class of attempts, generically classified as superunification theory, to explain the four known forces of nature—gravitational, electromagnetic, weak, and strong—on a single unifying basis. Common to all such schemes are the postulates that quantum mechanics and special relativity underlie the theoretical framework. Another common feature is supersymmetry, the notion that particles with half-integer values of the spin angular momentum (fermions) can be transformed into particles with integer spins (bosons).

The distinguishing feature of superstring theory is the postulate that elementary particles are not mere points in space but have linear extension. The characteristic linear dimension is given as a certain combination of the three most fundamental constants of nature: (1) Planck's constant h (named after the German physicist Max Planck, the founder of quantum physics), (2) the speed of light c, and (3) the universal gravitational constant G. The combination, called the Planck length $(Gh/c^3)^{1/2}$, equals roughly 10^{-33} cm, far smaller than the distances to which elementary particles can be probed in particle accelerators on Earth.

The energies needed to smash particles to within a Planck length of each other were available to the universe at a time equal to the Planck length divided by the speed of light. This time, called the Planck time $(Gh/c^5)^{1/2}$, equals approximately 10^{-43} second. At the Planck time, the mass density of the universe is thought to approach the Planck density, c^5/hG^2, roughly 10^{93} grams per cubic centimetre. Contained within a Planck volume is a Planck mass $(hc/G)^{1/2}$, roughly 10^{-5} gram. An object of such mass would be a quantum black hole, with an event horizon close to both its own Compton length (distance over which a particle is quantum mechanically "fuzzy") and the size of the cosmic horizon at the Planck time. Under such extreme conditions, space-time cannot be treated as a classical continuum and must be given a quantum interpretation.

The latter is the goal of the superstring theory, which has as one of its features the curious notion that the four space-time dimensions (three space dimensions plus one time dimension) of the familiar world may be an illusion. Real space-time, in accordance with this picture, has 26 or 10 space-time dimensions, but all of these dimensions except the usual four are somehow compacted or curled up to a size comparable to the Planck scale. Thus has the existence of these other dimensions escaped detection. It is presumably only during the Planck era, when the usual four space-time dimensions acquire their natural Planck scales, that the existence of what is more fundamental than the usual ideas of mass-energy and space-time becomes fully revealed. Unfortunately, attempts to deduce anything more quantitative or physically illuminating from the theory have bogged down in the intractable mathematics of this difficult subject. At the present time superstring theory remains more of an enigma than a solution.

Inflation

One of the more enduring contributions of particle physics to cosmology is the prediction of inflation by the American physicist Alan Guth and others. The basic idea is that at high energies matter is better described by fields than by classical means. The contribution of a field to the energy density (and therefore the mass density) and the pressure of the vacuum state need not have been zero in the past, even if it is today. During the time of superunification (Planck era, $10-43$ second) or grand unification (GUT era, $10-35$ second), the lowest-energy state for this field may have corresponded to a "false vacuum," with a combination of mass density and negative pressure that results gravitationally in a large repulsive force. In the context of Einstein's theory of general relativity, the false vacuum may be thought of alternatively as contributing a cosmological constant about 10100 times larger than it can possibly be today. The corresponding repulsive force causes the universe to inflate exponentially, doubling its size roughly once every 10^{-43} or 10^{-35} second. After at least 85 doublings, the temperature, which started out at 1032 or 1028 K, would have dropped to very low values near absolute zero. At low temperatures the true vacuum state may have lower energy than the false vacuum state, in an analogous fashion to how solid ice has lower energy than liquid water. The supercooling of the universe may therefore have induced a rapid phase transition from the false vacuum state to the true vacuum state, in which the cosmological constant is essentially zero. The transition would have released the energy differential (akin to the "latent heat" released by water when it freezes), which reheats the universe to high temperatures. From this temperature bath and the gravitational energy of expansion would then have emerged the particles and antiparticles of noninflationary big bang cosmologies.

Cosmic inflation serves a number of useful purposes. First, the drastic stretching during inflation flattens any initial space curvature, and so the universe after inflation will look exceedingly like an Einstein–de Sitter universe. Second, inflation so dilutes the concentration of any magnetic monopoles appearing as "topological knots" during the GUT era that their cosmological density will drop to negligibly small and acceptable values. Finally, inflation provides a mechanism for understanding the overall isotropy of the cosmic microwave background because the matter and radiation of the entire observable universe were in good thermal contact (within the cosmic event horizon) before inflation and therefore acquired the same thermodynamic characteristics. Rapid inflation carried different portions outside their individual event horizons. When inflation ended and the universe reheated and resumed normal expansion, these different portions, through the natural passage of time, reappeared on our horizon. And through the observed isotropy of the cosmic microwave background, they are inferred still to have the same temperatures. Finally, slight anisotropies in the cosmic microwave background occurred because of quantum fluctuations in the mass density. The amplitudes of these small (adiabatic) fluctuations remained independent of comoving scale during the period of inflation. Afterward they grew gravitationally by a constant factor until the recombination era. Cosmic microwave photons seen from the last scattering surface should therefore exhibit a scale-invariant spectrum of fluctuations, which is exactly what the Cosmic Background Explorer satellite observed.

How The Polarization of Light Influenced The Birth of The Universe.

As influential as inflation has been in guiding modern cosmological thought, it has not resolved all internal difficulties. The most serious concerns the problem of a "graceful exit." Unless the effective potential describing the effects of the inflationary field during the GUT era corresponds to an extremely gently rounded hill (from whose top the universe rolls slowly in the transition from the false vacuum to the true vacuum), the exit to normal expansion will generate so much turbulence and inhomogeneity (via violent collisions of "domain walls" that separate bubbles of true vacuum from regions of false vacuum) as to make inexplicable the small observed amplitudes for the anisotropy of the cosmic microwave background radiation. Arranging a tiny enough slope for the effective potential requires a degree of fine-tuning that most cosmologists find philosophically objectionable.

Steady State Theory And Other Alternative Cosmologies

Big bang cosmology, augmented by the ideas of inflation, remains the theory of choice among nearly all astronomers, but, apart from the difficulties discussed above, no consensus has been reached concerning the origin in the cosmic gas of fluctuations thought to produce the observed galaxies, clusters, and superclusters. Most astronomers would interpret these shortcomings as indications of the incompleteness of the development of the theory, but it is conceivable that major modifications are needed.

An early problem encountered by big bang theorists was an apparent large discrepancy between the Hubble time and other indicators of cosmic age. This discrepancy was resolved by revision of Hubble's original estimate for H0, which was about an order of magnitude too large owing to confusion between Population I and II variable stars and between H II regions and bright stars. However, the apparent difficulty motivated Bondi, Hoyle, and Gold to offer the alternative theory of steady state cosmology in 1948.

An overview of the 20th-century controversy concerning the origin of the universe: the big-bang model was disparaged by Fred Hoyle, but Edwin Hubble's findings supported it.

By that year, of course, the universe was known to be expanding; therefore, the only way to explain a constant (steady state) matter density was to postulate the continuous creation of matter to offset the attenuation caused by the cosmic expansion. This aspect was physically very unappealing to many people, who consciously or unconsciously preferred to have all creation completed in virtually one instant in the big bang. In the steady state theory the average age of matter in the universe is one-third the Hubble time, but any given galaxy could be older or younger than this mean value. Thus, the steady state theory had the virtue of making very specific predictions, and for this reason it was vulnerable to observational disproof.

The first blow was delivered by British astronomer Martin Ryle's counts of extragalactic radio sources during the 1950s and '60s. These counts involved the same methods discussed above for the star counts by Dutch astronomer Jacobus Kapteyn and the galaxy counts by Hubble except that radio telescopes were used. Ryle found more radio galaxies at large distances from Earth than can be explained under the assumption of a uniform spatial distribution no matter which cosmological model was assumed, including that of steady state. This seemed to imply that radio galaxies must evolve over time in the sense that there were more powerful sources in the past (and therefore observable at large distances) than there are at present. Such a situation contradicts a basic tenet of the steady state theory, which holds that all large-scale properties of the universe, including the population of any subclass of objects like radio galaxies, must be constant in time.

The second blow came in 1965 with the discovery of the cosmic microwave background radiation. Though it has few adherents today, the steady state theory is credited as having been a useful idea for the development of modern cosmological thought as it stimulated much work in the field.

At various times, other alternative theories have also been offered as challenges to the prevailing view of the origin of the universe in a hot big bang: the cold big bang theory (to account for galaxy formation), symmetric matter-antimatter cosmology (to avoid an asymmetry between matter and antimatter), variable G cosmology (to explain why the gravitational constant is so small), tired-light cosmology (to explain redshift), and the notion of shrinking atoms in a nonexpanding universe (to avoid the singularity of the big bang). The motivation behind these suggestions is, as indicated in the parenthetical comments, to remedy some perceived problem in the standard picture. Yet, in most cases, the cure offered is worse than the disease, and none of the mentioned alternatives has gained much of a following. The hot big bang theory has ascended to primacy because, unlike its many rivals, it attempts to address not isolated individual facts but a whole panoply of cosmological issues. And, although some sought-after results remain elusive, no glaring weakness has yet been uncovered.

*****

Life on Earth

Life, living matter and, as such, matter that shows certain attributes that include responsiveness, growth, metabolism, energy transformation, and reproduction. Although a noun, as with other defined entities, the word life might be better cast as a verb to reflect its essential status as a process. Life comprises individuals, living beings, assignable to groups (taxa). Each individual is composed of one or more minimal living units, called cells, and is capable of transformation of carbon-based and other compounds (metabolism), growth, and participation in reproductive acts. Life-forms present on Earth today have evolved from ancient common ancestors through the generation of hereditary variation and natural selection. Although some studies state that life may have begun as early as 4.1 billion years ago, it can be traced to fossils dated to 3.5–3.7 billion years ago, which is still only slightly younger than Earth, which gravitationally accreted into a planet about 4.5 billion years ago. But this is life as a whole. More than 99.9 percent of species that have ever lived are extinct. The several branches of science that reveal the common historical, functional, and chemical basis of the evolution of all life include electron microscopy, genetics, paleobiology (including paleontology), and molecular biology.

The phenomenon of life can be approached in several ways: life as it is known and studied on planet Earth; life imaginable in principle; and life, by hypothesis, that might exist elsewhere in the universe. As far as is known, life exists only on Earth. Most life-forms reside in a thin sphere that extends about 23 km (14 miles) from 3 km (2 miles) beneath the bottom of the ocean to the top of the troposphere (lower atmosphere); the relative thickness is comparable to a coat of paint on a rubber ball. An estimated 10–30 million distinguishable species currently inhabit this sphere of life, or biosphere.

165

Definitions of Life

Much is known about life from points of view reflected in the various biological, or "life," sciences. These include anatomy (the study of form at the visible level), ultrastructure (the study of form at the microscopic level), physiology (the study of function), molecular biology and biochemistry (the study of form and function at chemical levels), ecology (the study of the relations of organisms with their environments), taxonomy (the naming, identifying, and classifying of organisms), ethology (the study of animal behaviour), and sociobiology (the study of social behaviour). Specific sciences that participate in the study of life focus more narrowly on certain taxa or levels of observation—e.g., botany (the study of plants), lichenology (the study of lichens, leafy or crusty individuals composed of permanent associations between algae or photosynthetic bacteria and fungi), herpetology (the study of amphibians and reptiles), microbiology (the study of bacteria, yeast, and other unicellular fungi, archaea, protists, viruses), zoology (the study of marine and land animals), and cytology (the study of cells). Although the scientists, technicians, and others who participate in studies of life easily distinguish living matter from inert or dead matter, none can give a completely inclusive, concise definition of life itself. Part of the problem is that the core properties of life—growth, change, reproduction, active resistance to external perturbation, and evolution—involve transformation or the capacity for transformation. Living processes are thus antithetical to a desire for tidy classification or final definition. To take one example, the number of chemical elements involved with life has increased with time; an exhaustive list of the material constituents of life would therefore be premature. Nonetheless, most scientists implicitly use one or more of the metabolic, physiological, biochemical, genetic, thermodynamic, and autopoietic definitions given below.

Metabolic

Metabolic definitions are popular with biochemists and some biologists. Living systems are objects with definite boundaries, continually exchanging some materials with their surroundings but without altering their general properties, at least over some period of time. However, there are exceptions. There are frozen seeds and spores that remain, so far as is known, perfectly dormant. At low temperatures they lack metabolic activity for hundreds, perhaps thousands, of years but revive perfectly well upon being subjected to more clement conditions. A candle flame has a well-defined shape with a fixed boundary and is maintained by "metabolizing" its organic waxes and the surrounding molecular oxygen to produce carbon dioxide and water. Similar reactions, incidentally, occur in animals and plants. Flames also have a well-known capacity for growth. These facts underscore the inadequacy of this metabolic definition, even as they suggest the indispensable role of energy transformation to living systems.

Physiological

Physiological definitions of life are popular. Life is defined as any system capable of performing functions such as eating, metabolizing, excreting, breathing, moving, growing, reproducing, and responding to external stimuli. But many such properties are either present in machines that nobody is willing to call alive or absent from organisms, such as the dormant hard-covered seed of a tree, that everybody is willing to call alive. An automobile, for example, can be said to eat, metabolize, excrete, breathe, move, and be responsive to external stimuli. A visitor from another planet, judging from the enormous number of automobiles on Earth and the way in which cities and landscapes have been designed for the special benefit of motorcars, might well believe that automobiles are not only alive but are the dominant life-form on the planet.

Biochemical

A biochemical or molecular biological definition sees living organisms as systems that contain reproducible hereditary information coded in nucleic acid molecules and that metabolize by controlling the rate of chemical reactions using the proteinaceous catalysts known as enzymes. In many respects, this is more satisfying than the physiological or metabolic definitions of life. However, even here there are counterexamples. Viruslike agents called prions lack nucleic acids, although the nucleic acids of the animal cells in which they reside may be involved in their reproduction. Ribonucleic acid (RNA) molecules may replicate, mutate, and then replicate their mutations in test tubes, although by themselves they are not alive. Furthermore, a definition strictly in chemical terms seems peculiarly vulnerable. It implies that, were a person able to construct a system that had all the functional properties of life, it would still not be alive if it lacked the molecules that earthly biologists are fond of—and made of.

Genetic

All organisms on Earth, from the tiniest cell to the loftiest trees, display extraordinary powers. They effortlessly perform complex transformations of organic molecules, exhibit elaborate behaviour patterns, and indefinitely construct from raw materials in the environment more or less identical copies of themselves. How could systems of such staggering complexity and such stunning beauty ever arise? A main part of the answer, for which today there is excellent scientific evidence, was carefully chronicled by the English naturalist Charles Darwin in the years before the publication in 1859 of his epoch-making work On the Origin of Species. A modern rephrasing of his theory of natural selection goes something like this: Hereditary information is carried by large molecules known as genes, composed of nucleic acids. Different genes are responsible for the expression of different characteristics of the organism. During the reproduction of the organism, the genes also replicate and thereby pass on the instructions for various characteristics to the next generation. Occasionally, there are imperfections, called mutations, in gene replication. A mutation alters the instructions for one or more particular characteristics. The mutation also breeds true, in the sense that its capability for determining a given characteristic of the organism remains unimpaired for generations until the mutated gene is itself mutated. Some mutations, when expressed, will produce characteristics favourable for the organism; organisms with such favourable genes will reproduce preferentially over those without such genes. Most mutations, however, turn out to be deleterious and often lead to some impairment or to death of the organism. (To illustrate, it is unlikely that one can improve the functioning of a finely crafted watch by dropping it from a tall building. The watch may perform better, but this is highly improbable.) In this way, organisms slowly evolve toward greater complexity. This evolution occurs, however, only at enormous cost: modern humans, complex and reasonably well-adapted, exist only because of billions of deaths of organisms slightly less adapted and somewhat less complex. In short, Darwin's theory of natural selection states that complex organisms evolved through time because of replication, mutation, and replication of mutations. A genetic definition of life therefore would be a system capable of evolution by natural selection.

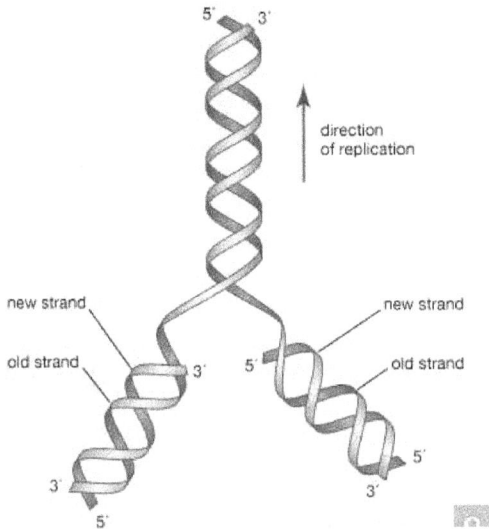

The initial proposal of the structure of DNA by James Watson and Francis Crick, which was accompanied by a suggestion on the means of replication.

This definition places great emphasis on the importance of replication. Replication refers to the capacity of molecules such as deoxyribonucleic acid (DNA) to precisely copy themselves, whereas reproduction refers to the increase in number of organisms by acts that make a new individual from its parent or parents. In any organism, enormous biological effort is directed toward reproduction, although it confers no obvious benefit on the reproducing organism itself. However, if life is defined as an entity capable of reproduction, then a mule, which is clearly alive yet does not reproduce, would be excluded from the living under this restrictive definition. Indeed, many organisms, such as hybrid mammals and plants that are past their prime, do not reproduce even though the individual cells of which they are composed may.

Life defined as a reproductive system dependent on replicating components does not rule out synthetic reproduction. For example, it should be possible to construct a machine that is capable of producing identical copies of itself from preformed building blocks but that arranges its descendants in a slightly different manner when a random change occurs in its instructions. Such a machine would of course reproduce its instructions as well. But the fact that such a machine would satisfy the genetic definition of life is not an argument against such a definition; in fact, if the building blocks were simple enough, the machine would have the capability of evolving into very complex systems that would probably have all the other properties attributed to living systems. (Some computer programmers have already claimed, on the basis of running generations of replicating and mutating computer instructions, to have created artificial life ["a-life"]; such programs do not, however, show any real freedom or awareness, and their activities are thus far limited to the insides of computers.) The genetic definition has the additional advantage of being expressed purely in functional terms; i.e., it does not depend on any particular choice of constituent molecules. The improbability of contemporary organisms—dealt with more fully below—is so great that these organisms could not possibly have arisen by purely random processes and without historical continuity. Fundamental to the genetic definition of life then seems to be the notion that a certain level of complexity cannot be achieved without natural selection.

Thermodynamic
Thermodynamics distinguishes between isolated, closed, and open systems. An isolated system is separated from the rest of the environment and exchanges neither light nor heat nor matter with its surroundings. A closed system exchanges energy but not matter. An open system is one in which both material and energetic exchanges occur. The second law of thermodynamics states that, in a closed system, no processes will tend to occur that increase the net organization (or decrease the net entropy) of the system. Thus, the universe taken as a whole is steadily moving toward a state of complete randomness, lacking any order, pattern, or beauty. This fate was popularized in the 19th century as the "heat death" of the universe.

Living organisms are manifestly organized and at first sight seem to represent a contradiction to the second law of thermodynamics. Indeed, living systems might then be defined as localized regions where there is a continuous maintenance or increase in organization. Living systems, however, do not really contradict the second law. They increase their organization in regions of energy flow, and, indeed, their cycling of materials and their tendency to grow can be understood only in the context of a more general definition of the second law that applies to open as well as closed and isolated systems. In nature (except at cosmic scales, where gravity becomes a crucial factor), energy moves from being concentrated to being spread out; spontaneously occurring complex systems do not violate the second law but help energy spread out, thus producing entropy and reducing gradients.

A general statement of open-system thermodynamics is that nature abhors a gradient, a difference across a distance. Differences and gradients in nature represent improbable, preexisting organizations. Many complex systems in nature spontaneously arise to degrade gradients and persist until the gradients are nullified. A tornado, for example, is an improbable, matter-cycling system that appears in the area of a barometric pressure gradient; when the air pressure gradient is gone, the "improbable" tornado disappears. Life seems to be a similar system, but one that degrades the solar gradient, the electromagnetic difference between the extremely hot (5,800 K [5,500 °C, or 10,000 °F]) Sun and very cold (2.7 K [−270.3 °C, or −454.5 °F]) outer space. (K = kelvin. On the Kelvin temperature scale, in which 0 K [−273 °C, or −460 °F] is absolute zero, 273 K [0 °C, or 32 °F] is the freezing point of water, and 373 K [100 °C, or 212 °F] is the boiling point of water at one atmosphere pressure.) Most life on Earth is dependent on the flow of sunlight, which is utilized by photosynthetic organisms to construct complex molecules from simpler ones. Some deep-sea and cave organisms called chemoautotrophs depend on chemical gradients, such as the natural energy-producing reaction between hydrogen sulfide bubbling up from vents and oxygen dissolved in water. The organization of life on Earth can thus be seen as being driven by a natural second-law-based reduction between the energy of the hot Sun and the cooler space around it. Although life has not fully reduced the solar gradient, incorporation of carbon dioxide into chemoautotrophs and production of clouds by plants help keep Earth's surface cooler than it would otherwise be, thereby helping to degrade the solar energy gradient.

Some scientists argue on grounds of quite general open-system thermodynamics that the organization of a system increases as energy flows through it. Moreover, energy flow leads to the development of cycles. An example of a biological cycle on Earth is the carbon cycle. Carbon from atmospheric carbon dioxide is incorporated by photosynthetic or chemosynthetic organisms and converted into carbohydrates through the process of autotrophy. These carbohydrates are ultimately oxidized by heterotrophic organisms to extract useful energy locked in their chemical bonds. In the oxidation of carbohydrates, carbon dioxide is returned to the atmosphere, thus completing the cycle. It has been shown that similar cycles develop spontaneously and in the absence of life by the flow of energy through chemical systems. Biological cycles may represent natural thermodynamic cycles reinforced by a genetic apparatus. It seems doubtful that open-system thermodynamic processes in the absence of replication lead to the sorts of complexity that characterize biological systems; replication, however, may be interpreted as an especially efficient thermodynamic means of gradient breakdown—a kind of special, slow-burning "fire." In any case, it is clear that much of the complexity of life on Earth has arisen through replication, with thermodynamically favoured pathways being used by energy-transforming organisms.

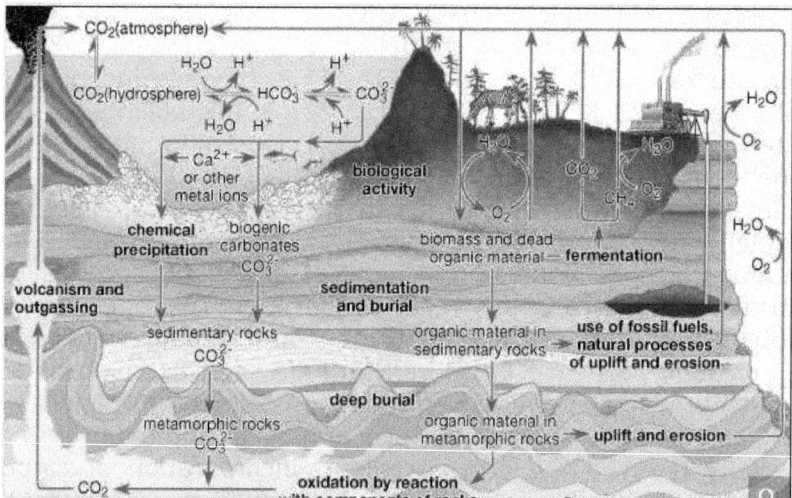

The carbon cycleCarbon is transported in various forms through the atmosphere, the hydrosphere, and geologic formations. One of the primary pathways for the exchange of carbon dioxide (CO_2) takes place between the atmosphere and the oceans; there a fraction of the CO_2 combines with water, forming carbonic acid (H_2CO_3) that subsequently loses hydrogen ions (H+) to form bicarbonate (HCO3−) and carbonate (CO32−) ions. Mollusk shells or mineral precipitates that form by the reaction of calcium or other metal ions with carbonate may become buried in geologic strata and eventually release CO2 through volcanic outgassing. Carbon dioxide also exchanges through photosynthesis in plants and through respiration in animals. Dead and decaying organic matter may ferment and release CO2 or methane (CH4) or may be incorporated into sedimentary rock, where it is converted to fossil fuels. Burning of hydrocarbon fuels returns CO2 and water (H2O) to the atmosphere. The biological and anthropogenic pathways are much faster than the geochemical pathways and, consequently, have a greater impact on the composition and temperature of the atmosphere.

Autopoietic

A newer definition of life revolves around the idea of autopoiesis. This idea was put forth by Chilean biologists Humberto Maturana and Francisco Varela and emphasizes the peculiar closure of living systems, which are alive and maintain themselves metabolically whether they succeed in reproduction or not. Unlike machines, whose governing functions are embedded by human designers, organisms are self-governing. The autopoietic definition of life resembles the physiological definition but emphasizes life's maintenance of its own identity, its informational closure, its cybernetic self-relatedness, and its ability to make more of itself. Autopoiesis refers to self-producing, self-maintaining, self-repairing, and self-relational aspects of living systems. Living beings maintain their form by the continuous interchange and flow of chemical components. Cellular autopoietic systems are bounded by a dynamic material made by the system itself. In life on Earth the limiting material is lipid membrane studded with transport proteins fabricated by the incessantly active cell. A source of usable energy flows to all living or autopoietic systems—either light in the visible or near-visible spectrum or specific organic carbon or other chemicals such as hydrogen, hydrogen sulfide, or ammonia. Energy sources that have never been adequate to maintain autopoiesis on Earth include heat, sound waves, and electromagnetic radiation outside the visible or near-visible spectrum.

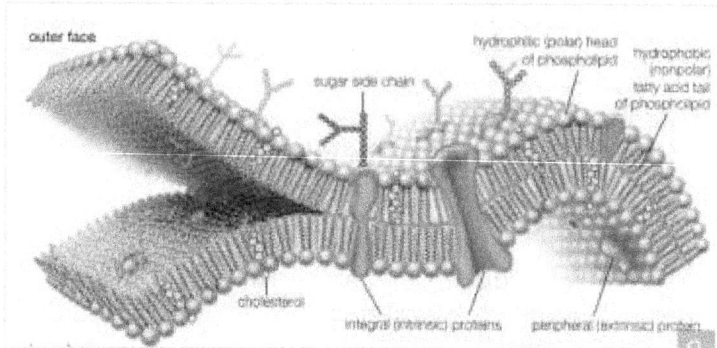

Molecular View of the Cell Membrane

Intrinsic proteins penetrate and bind tightly to the lipid bilayer, which is made up largely of phospholipids and cholesterol and which typically is between 4 and 10 nanometers (nm; 1 nm = 10^{-9} metre) in thickness. Extrinsic proteins are loosely bound to the hydrophilic (polar) surfaces, which face the watery medium both inside and outside the cell. Some intrinsic proteins present sugar side chains on the cell's outer surface.

One of the difficulties in defining life is that the only example is life found on the third planet from the Sun. On Earth all life's autopoietic systems require a supply of water in its liquid state for self-maintenance of their parts. Taken together, all transformations that underlie autopoiesis require six elements: carbon, nitrogen, hydrogen, oxygen, phosphorus, and sulfur. The chemical components of all living entities are fashioned primarily from these elements.

The smallest autopoietic system on Earth is a living bacterial cell. (Viruses, plasmids, and other replicating molecules cannot, even in principle, behave as an autopoietic system; no matter how much food, liquid water, and serviceable energy they are provided, they still require cells for their continuity and duplication.) Some cells, such as Carsonella ruddii, have fewer than 200 genes and proteins, but they, like organelles and viruses, are not autopoietic, since they must be inside an autopoietic system (living cell) to metabolize and survive. No self-bounded autopoietic system smaller than a cell with at least 450 proteins and the genes that code for these proteins has ever been described. Larger than bacteria are other autopoietic systems of intermediate size such as protists, fungal spores, mules and other individual mammals, and plants such as oak trees or poppies. Autopoietic entities at even larger levels include ecosystems such as coral reefs, prairies, or ponds. The maximal or largest single autopoietic system known is referred to as "Gaia," named by English atmospheric scientist James E. Lovelock for Gaea, the ancient Greek personification of Earth. Gaia is basically a closed thermodynamic system because there is little interchange of matter with the extraterrestrial environment. There is evidence that the global, Gaian system of life shows organism-like properties, such as regulation of atmospheric chemistry, global mean temperature, and oceanic salinity over multimillion-year time spans. Such regulation may be understood as part of life's organization as a complex and cyclical open thermodynamic system.

The existence of diverse definitions of life, as detailed in the previous section, surely means that life is complex and difficult to briefly define. A scientific understanding of living systems has existed since the second half of the 19th century. But the diversity of definitions and lack of consensus among professionals suggest something else as well. As detailed in this section, all organisms on Earth are extremely closely related, despite superficial differences. The fundamental pattern, both in form and in matter, of all life on Earth is essentially identical. Also, as noted in this section, this identity implies that all organisms on Earth are evolved from a single instance of the origin of life. To generalize from a single example is difficult, especially when the example itself is changing, growing, and evolving. In this respect the biologist is fundamentally handicapped, as compared with, say, the chemist, physicist, geologist, or meteorologist, each of whom can now study aspects of his discipline beyond Earth. If truly only one sort of life on Earth exists, then perspective is lacking in a most fundamental way. On the other hand, the historical continuity of all life-forms means that ancient life, perhaps even the origins of life, may be glimpsed by studying modern cells.

millions of years ago

| 4,000 | 2,500 | 541 | 500 | 450 |

| Hadean | Archean | Proterozoic | Paleozoic |

| Eoarchean | Mesoarchean | Paleoproterozoic | Neoproterozoic | Cambrian | Ordovician | Silurian |
| Paleoarchean | Neoarchean | Mesoproterozoic | | | |

first living organisms-bacteria

first animal traces and algae

algae

jellyfish

first land plants

fishes

single-celled living things

first animals with shells

trilobites

first fishes

corals

multicellular living things

sponges

many types of animals with shells

Precambrian Eon

Phanerozoic Eon

millions of years ago

| 400 | 350 | 300 | 250 | 200 | 150 | 100 | 50 | 2.6 | 0 |

| Paleozoic | Mesozoic | Cenozoic |

| Devonian | Carboniferous | Permian | Triassic | Jurassic | Cretaceous | Paleogene | Quaternary |
| | | | | | | Neogene | |

treelike ferns

fishes

first dinosaurs

dinosaurs

many types of mammals

first insects

many types of reptiles

conifers

palmlike plants

first flowering plants

humans

corals

first reptiles

first mammals

first birds

first amphibians

ocean-living reptiles

mammoths

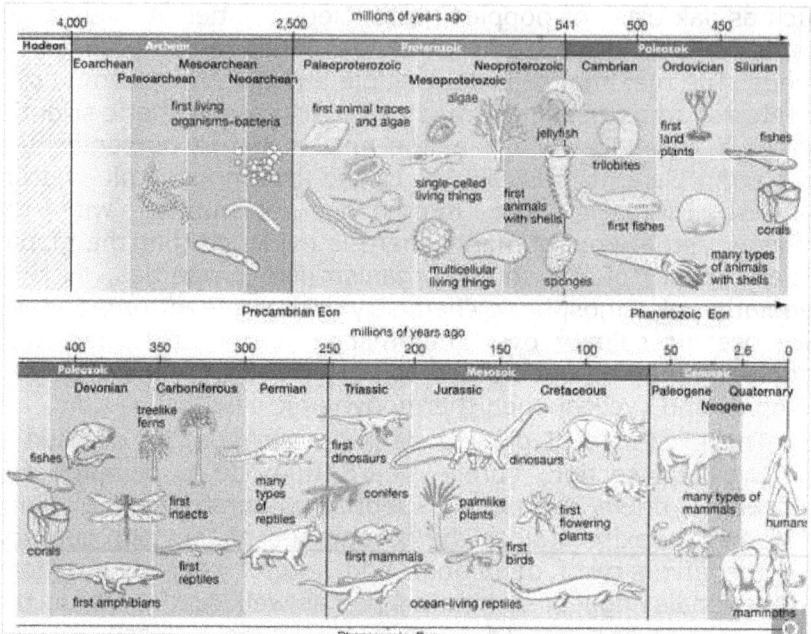

Over hundreds of millions of years, life spread through the seas and over Earth's surface. The first life-forms were small and simple. Later forms were more complicated and diverse.

The Biosphere

Reconnaissance missions to the planets of the inner solar system have revealed stark and barren landscapes. From the heavily cratered and atmosphereless surfaces of both Mercury and the Moon to the hot sulfurous fogs of Venus and the dusty, windswept surface of Mars, no sign of life is apparent anywhere. The biosphere, by definition the place where all Earth's life dwells, is a delight with its green, wet contrast. Austrian geologist Eduard Suess invented the term biosphere to match the other envelopes of the planet: the atmosphere of gas; the hydrosphere of oceans, lakes, rivers, springs and other waters; and the lithosphere, or the solid rock surface of the outer portion of Earth. Yet it was the great Russian crystallographer and mineralogist Vladimir I. Vernadsky who brought the term into common parlance with his book of the same name. In The Biosphere (1926) Vernadsky outlines his view of life as a major geological force. Living matter, Vernadsky contends, erodes, levels, transports, and chemically transforms surface rocks, minerals, and other features of Earth. If the biosphere is the place where life is found, the biota (or the biomass as a whole) is the sum of all living forms: flora, fauna, and microbiota.

During the second half of the 20th century, study of the deep sea, the upper atmosphere, the Antarctic dry deserts, newly opened caves, sulfurous tunnels, and granitic rocks showed that Earth's surface is vigorously inhabited in places that were unknown to Vernadsky and his colleagues. Vernadsky's international school of thought ushered in the field of "biogeochemistry," and chemists and geologists were recruited to consider life as a planetary phenomenon. But not until giant, mouthless, red-gilled tube worms were videographed in the late 1970s and '80s did the extent and the weirdness of Earth's biota begin to be fathomed. Entire large ecosystems were recognized on the ocean's bottom that live not by the usual plant photosynthesis but rather by chemolithoautotrophy, a kind of metabolism in which organisms make food from carbon dioxide using energy from the oxidation of sulfide, methane, or other inorganic compounds. These discoveries have led to a deeper understanding of life's varied modes of nutrition and sources of energy. Bacterial symbionts living in the tissues of some polychaete worms (alvinellids) or pogonophora (such as Riftia pacytila) provide the animals with their total nutritional needs. The submarine ecosystems supported by bacteria thrive along the worldwide rift zones that extend along the borders of huge continental plates at the Mid-Atlantic Ridge, on the East Pacific Rise, at 21° north of the Equator off the coast of Baja California, Mex., and at a dozen other newly studied sites. By the beginning of the 21st century it had become abundantly clear that many life-forms and ecosystems remained unknown or under-studied. Those in the Siberian tundra, in the thickly forested portions of the Amazon River valley and its tributaries, at the tops of remote mountains and inside granitic rocks in temperate zones, and in the centre of Africa remain as inaccessible to most naturalists as they have been throughout history. The easily accessed woodlands and fields of well-lit land surfaces are another story.

On land, 24 percent of the productivity of organic carbon biomass generated by plants is directly controlled by burgeoning populations of one species, humans. As Vernadsky noted, life in general and human life in particular tend to accelerate the number of materials and the rate of flow of these materials through the biosphere, the place where all life exists—so far.

Vernadsky anticipated new discoveries of life inside hot springs and granitic rock. Although he qualified this statement by asserting that it would not hold for temporary, abnormal circumstances, such as during a lava flow or a volcanic eruption, he wrote,

Thus far, we have seen that the biosphere, by structure, composition, and physical makeup, is completely enclosed by the domain of life, which has so adapted itself to biospheric conditions that there is no place [on Earth] in which it is unable to manifest itself in one way or another.

Although much is not known about life in the depths of the rocks and the sea, determination of the total range and mass (biomass) of the biota, the sum of all life in the biosphere, is a reasonable scientific goal.

Chemistry of Life

Human beings, like mammals in general, are ambulatory collections of some 1014 cells. Human cells are in all fundamental respects the same as those that make up the other animals. Each cell typically consists of one central, spherical nucleus and another heterogeneous region, the cytoplasm. (Only bacterial cells lack nuclei; those of plants, fungi, and all other organisms contain one or more nuclei.) A living nucleated cell, a marvel of detailed and complex architecture, appears frenetic with activity when seen through a microscope. On a deeper chemical level, it is known that life's large molecules, the proteins and nucleic acids, are synthesized at a very fast rate. Enzymes, which speed up chemical reactions, are all proteins, but by no means are all proteins enzymes. An enzyme catalyzes the synthesis of more than 1,000 other molecules per second. The total mass of a metabolizing bacterial cell can be synthesized in 20 minutes. The information content of a small cell has been estimated as about 10^{10} bits, comparable to about 10^6 (or one million) pages of the print version of an Encyclopædia. Although some feel debased by the implication that people are "nothing more" than a frenetic collection of interacting molecules, others are thrilled with the power of science to reveal the inner workings of the chemistry of life. The spectacular success of biochemistry and molecular biology in the 20th century suggests that laws of biology are derived from the interaction of atoms, thermodynamic principles, and life's chemistry, which has persisted with faithful continuity since its origin some 3.7 billion to 3.5 billion years ago.

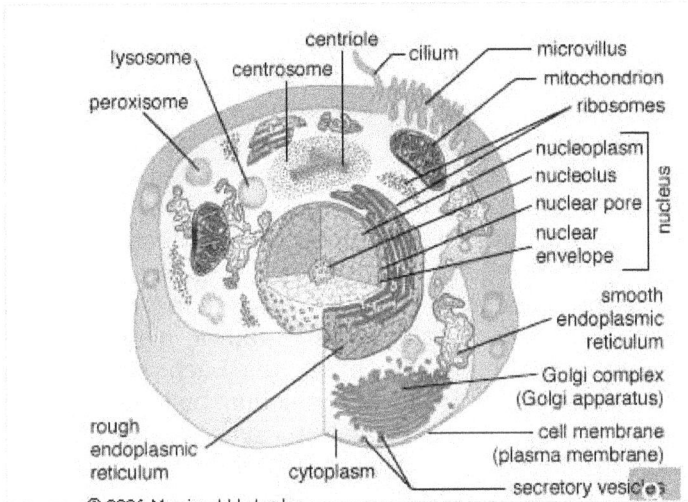

Animal Cell

Principal structures of an animal cellCytoplasm surrounds the cell's specialized structures, or organelles. Ribosomes, the sites of protein synthesis, are found free in the cytoplasm or attached to the endoplasmic reticulum, through which materials are transported throughout the cell. Energy needed by the cell is released by the mitochondria. The Golgi complex, stacks of flattened sacs, processes and packages materials to be released from the cell in secretory vesicles. Digestive enzymes are contained in lysosomes. Peroxisomes contain enzymes that detoxify dangerous substances. The centrosome contains the centrioles, which play a role in cell division. The microvilli are fingerlike extensions found on certain cells. Cilia, hairlike structures that extend from the surface of many cells, can create movement of surrounding fluid. The nuclear envelope, a double membrane surrounding the nucleus, contains pores that control the movement of substances into and out of the nucleoplasm. Chromatin, a combination of DNA and proteins that coil into chromosomes, makes up much of the nucleoplasm. The dense nucleolus is the site of ribosome production.

DNA, RNA, and Protein

The specific carrier of the genetic information in all organisms is the nucleic acid known as DNA, short for deoxyribonucleic acid. DNA is a double helix, two molecular coils wrapped around each other and chemically bound one to another by bonds connecting adjacent bases. Each long ladderlike DNA helix has a backbone that consists of a sequence of alternating sugars and phosphates. Attached to each sugar is a "base" consisting of the nitrogen-containing compound adenine, guanine, ctyosine, or thymine. Each sugar-phosphate-base "rung" is called a nucleotide. A very significant one-to-one pairing between bases occurs that ensures the connection of adjacent helices. Once the sequence of bases along one helix (half the ladder) has been specified, the sequence along the other half is also specified. The specificity of base pairing plays a key role in the replication of the DNA molecule. Each helix makes an identical copy of the other from molecular building blocks in the cell. These nucleic acid replication events are mediated by enzymes called DNA polymerases. With the aid of enzymes, DNA can be produced in the laboratory.

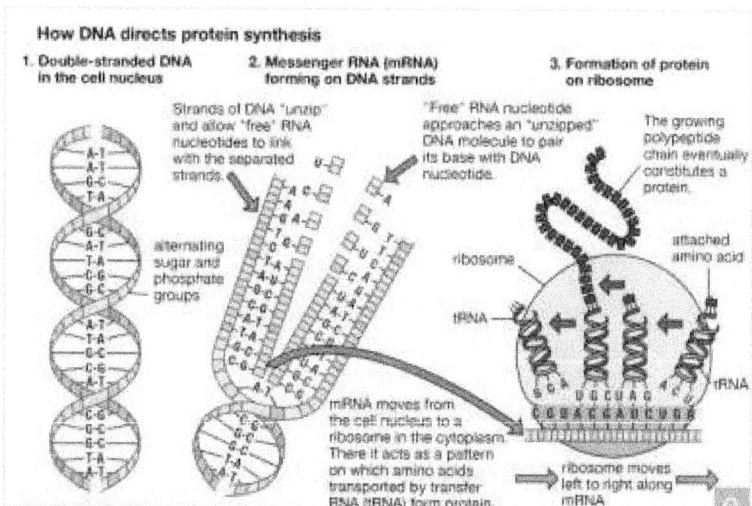

DNA in the cell nucleus carries a genetic code, which consists of sequences of adenine (A), thymine (T), guanine (G), and cytosine (C) (Figure 1). RNA, which contains uracil (U) instead of thymine, carries the code to protein-making sites in the cell. To make RNA, DNA pairs its bases with those of the "free" nucleotides (Figure 2). Messenger RNA (mRNA) then travels to the ribosomes in the cell cytoplasm, where protein synthesis occurs (Figure 3). The base triplets of transfer RNA (tRNA) pair with those of mRNA and at the same time deposit their amino acids on the growing protein chain. Finally, the synthesized protein is released to perform its task in the cell or elsewhere in the body.

The cell, whether bacterial or nucleated, is the minimal unit of life. Many of the fundamental properties of cells are a function of their nucleic acids, their proteins, and the interactions among these molecules bounded by active membranes. Within the nuclear regions of cells is a mélange of twisted and interwoven fine threads, the chromosomes. Chromosomes by weight are composed of 50–60 percent protein and 40–50 percent DNA. During cell division, in all cells but those of bacteria (and some ancestral protists), the chromosomes display an elegantly choreographed movement, separating so that each offspring of the original cell receives an equal complement of chromosomal material. This pattern of segregation corresponds in all details to the theoretically predicted pattern of segregation of the genetic material implied by the fundamental genetic laws. The chromosome combination of the DNA and the proteins (histone or protamine) is called nucleoprotein. The DNA stripped of its protein is known to carry genetic information and to determine details of proteins produced in the cytoplasm of cells; the proteins in nucleoprotein regulate the shape, behaviour, and activities of the chromosomes themselves.

The other major nucleic acid is ribonucleic acid (RNA). Its five-carbon sugar is slightly different from that of DNA. Thymine, one of the four bases that make up DNA, is replaced in RNA by the base uracil. RNA appears in a single-stranded form rather than a double. Proteins (including all enzymes), DNA, and RNA have a curiously interconnected relation that appears ubiquitous in all organisms on Earth today. RNA, which can replicate itself as well as code for protein, may be older than DNA in the history of life.

Chemistry in Common

DNA triplet	RNA triplet	amino acid	DNA triplet	RNA triplet	amino acid
AAA	UUU	phenylalanine	ACA	UGU	cysteine
AAG	UUC		ACG	UGC	
AAT	UUA		ACC	UGG	tryptophan
AAC	UUG				
GAA	CUU	leucine	ATA	UAU	tyrosine
GAG	CUC		ATG	UAC	
GAT	CUA				
GAC	CUG		ATT	UAA	(termination: end of specification)
			ATC	UAG	
AGA	UCU		ACT	UGA	
AGG	UCC				
AGT	UCA	serine	GCA	CGU	arginine
AGC	UCG		GCG	CGC	
TCA	AGU		GCT	CGA	
TCG	AGC		GCC	CGG	
			TCT	AGA	
GGA	CCU	proline	TCC	AGG	
GGG	CCC				
GGT	CCA		GTA	CAU	histidine
GGC	CCG		GTG	CAC	
TAA	AUU	isoleucine (Ileu)	GTT	CAA	glutamine (GluN)
TAG	AUC		GTC	CAG	
TAT	AUA				
			TTA	AAU	asparagine (AspN)
TAC	AUG	methionine	TTG	AAC	
TGA	ACU	threonine	TTT	AAA	lysine
TGG	ACC		TTC	AAG	
TGT	ACA				
TGC	ACG		CCA	GGU	glycine
			CCG	GGC	
CAA	GUU	valine	CCT	GGA	
CAG	GUC		CCC	GGG	
CAT	GUA				
CAC	GUG		CTA	GAU	aspartic acid
			CTG	GAC	
CGA	GCU	alanine	CTT	GAA	glutamic acid
CGG	GCC		CTC	GAG	
CGT	GCA				
CGC	GCG				

*The columns may be read: the DNA triplet is transcribed into an RNA triplet, which then directs the production of an amino acid.

The Genetic Code: Nucleotide Triplets (Codons)
Specifying Different Amino Acids in Protein Chains

The genetic code was first broken in the 1960s. Three consecutive nucleotides (base-sugar-phosphate rungs) are the code for one amino acid of a protein molecule. By controlling the synthesis of enzymes, DNA controls the functioning of the cell. Of the four different bases taken three at a time, there are 43, or 64, possible combinations. The meaning of each of these combinations, or codons, is known. Most of them represent one of the 20 particular amino acids found in protein. A few of them represent punctuation marks—for example, instructions to start or stop protein synthesis. Some of the code is called degenerate. This term refers to the fact that more than one nucleotide triplet may specify a given amino acid. This nucleic acid–protein interaction underlies living processes in all organisms on Earth today. Not only are these processes the same in all cells of all organisms, but even the particular "dictionary" that is used for the transcription of DNA information into protein information is essentially the same. Moreover, this code has various chemical advantages over other conceivable codes. The complexity, ubiquity, and advantages argue that the present interactions among proteins and nucleic acids are themselves the product of a long evolutionary history. They must interact as a single reproductive, autopoietic system that has not failed since its origin. The complexity reflects time during which natural selection could accrue variations; the ubiquity reflects a reproductive diaspora from a common genetic source; and the advantages, such as the limited number of codons, may reflect an elegance born of use. DNA's "staircase" structure allows for easy increases in length. At the time of the origin of life, this complex replication and transcription apparatus could not have been in operation. A fundamental problem in the origin of life is the question of the origin and early evolution of the genetic code.

Many other commonalities exist among organisms on Earth. Only one class of molecules stores energy for biological processes until the cell has use for it; these molecules are all nucleotide phosphates. The most common example is adenosine triphosphate (ATP). For the very different function of energy storage, a molecule identical to one of the building blocks of the nucleic acids (both DNA and RNA) is employed. Metabolically ubiquitous molecules—flavin adenine dinucleotide (FAD) and coenzyme A—include subunits similar to the nucleotide phosphates. Nitrogen-rich ring compounds, called porphyrins, represent another category of molecules; they are smaller than proteins and nucleic acids and common in cells. Porphyrins are the chemical bases of the heme in hemoglobin, which carries oxygen molecules through the bloodstream of animals and the nodules of leguminous plants. Chlorophyll, the fundamental molecule mediating light absorption during photosynthesis in plants and bacteria, is also a porphyrin. In all organisms on Earth, many biological molecules have the same "handedness" (these molecules can have both "left-" and "right-handed" forms that are mirror images of each other; see below The earliest living systems). Of the billions of possible organic compounds, fewer than 1,500 are employed by contemporary life on Earth, and these are constructed from fewer than 50 simple molecular building blocks.

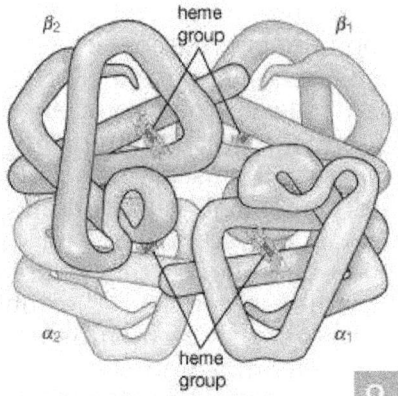

The hemoglobin tetramerTwo αβ dimers combine to form the complete hemoglobin molecule. Each heme group contains a central iron atom, which is available to bind a molecule of oxygen. The α1β2 region is the area where the α1 subunit interacts with the $β_2$ subunit.

Besides chemistry, cellular life has certain supramolecular structures in common. Organisms as diverse as single-celled paramecia and multicellular pandas (in their sperm tails), for example, possess little whiplike appendages called cilia (or flagella, a term that is also used for completely unrelated bacterial structures; the correct generic term is undulipodia). These "moving cell hairs" are used to propel the cells through liquid. The cross-sectional structure of undulipodia shows nine pairs of peripheral tubes and one pair of internal tubes made of proteins called microtubules. These tubules are made of the same protein as that in the mitotic spindle, the structure to which chromosomes are attached in cell division. There is no immediately obvious selective advantage of the 9:1 ratio.

Rather, these commonalities indicate that a few functional patterns based on common chemistry are used over and over again by the living cell. The underlying relations, particularly where no obvious selective advantage exists, show all organisms on Earth are related and descended from a very few common cellular ancestors—or perhaps one.

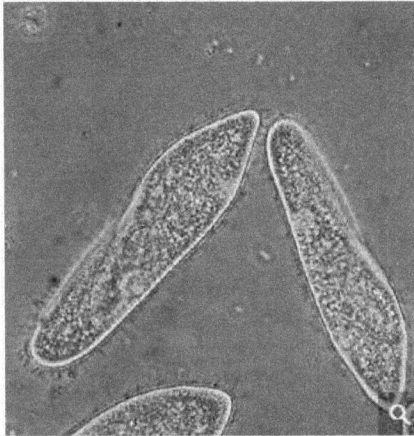

Paramecium caudatum.

Modes of Nutrition and Energy Generation

Chemical bonds that make up the compounds of living organisms have a certain probability of spontaneous breakage. Accordingly, mechanisms exist that repair this damage or replace the broken molecules. Furthermore, the meticulous control that cells exercise over their internal activities requires the continued synthesis of new molecules. Processes of synthesis and breakdown of the molecular components of cells are collectively termed metabolism. For synthesis to keep ahead of the thermodynamic tendencies toward breakdown, energy must be continuously supplied to the living system.

Energy, Carbon, and Electrons

Organisms acquire energy by two general methods: by light or by chemical oxidation. Productive organisms, called autotrophs, convert light or chemicals into energy-rich organic compounds beginning with energy-poor carbon dioxide (CO_2). These autotrophs provide energy for the other organisms, the heterotrophs. Heterotrophs are organisms that acquire their energy by the controlled breakdown of preexisting organic molecules, or food. Human beings, like most other animals, fungi, protists, and bacteria, are heterotrophs.

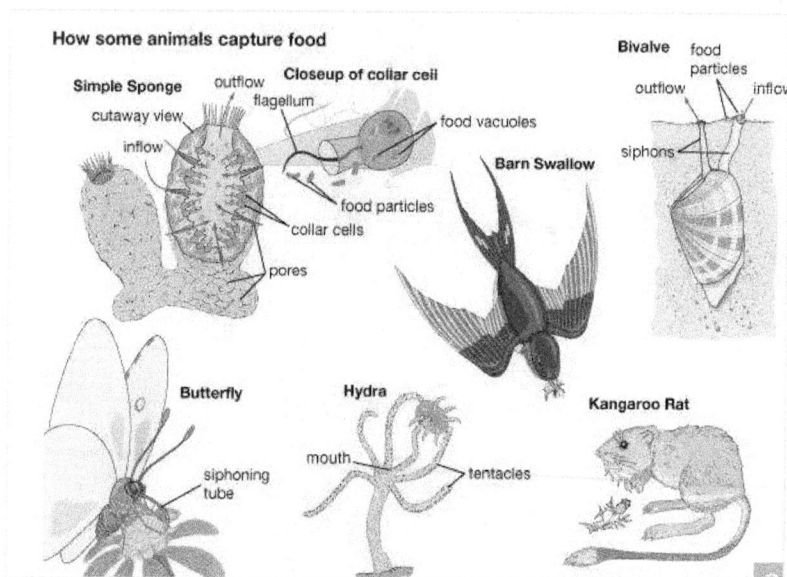

Food Capture: Animals

The Process of Food Capture in Different Animal Species

Autotrophic organisms are often primary producers in their ecosystems. They acquire their useful free energy from sources other than food: either from the energy of sunlight (photoautotrophs) or from oxidative chemical reactions (chemoautotrophs). The latter mode of metabolism refers to life-forms that use inorganic materials (ammonia [NH_3], methane [CH_4], or hydrogen sulfide [H_2S]) combined with oxygen to generate their energy. Only some bacteria are capable of obtaining energy by "burning" inorganic chemicals.

Green plants are typical photoautotrophs. Plants absorb sunlight to generate ATP and to disassociate water into oxygen and hydrogen. To break down the water molecule, H_2O, into hydrogen and oxygen requires much energy. The hydrogen from water is then combined in the "dark reactions" with carbon dioxide, CO_2. The result is the production of such energy-rich organic molecules as sugars, amino acids, and nucleotides. The oxygen becomes the gas O_2, which is released as waste back into the atmosphere. Animals, which are strictly heterotrophs, cannot live on carbon dioxide, sunlight, and water with a few salts like plants do. They must breathe in the atmospheric oxygen. Animals combine oxygen chemically with hydrogen atoms that they remove from their food—that is, from organic materials such as sugar, protein, and amino acids. Animals release water as a waste product from the oxygen respiration. Animals, like all heterotrophs, use organic materials as their sole source of carbon. This conversion of carbon provides an example of an aspect of an ecological cycle in which a required element flows through different types of organisms as it changes its oxidation state from CO_2 to $(CH_2O)n$ and back to CO_2.

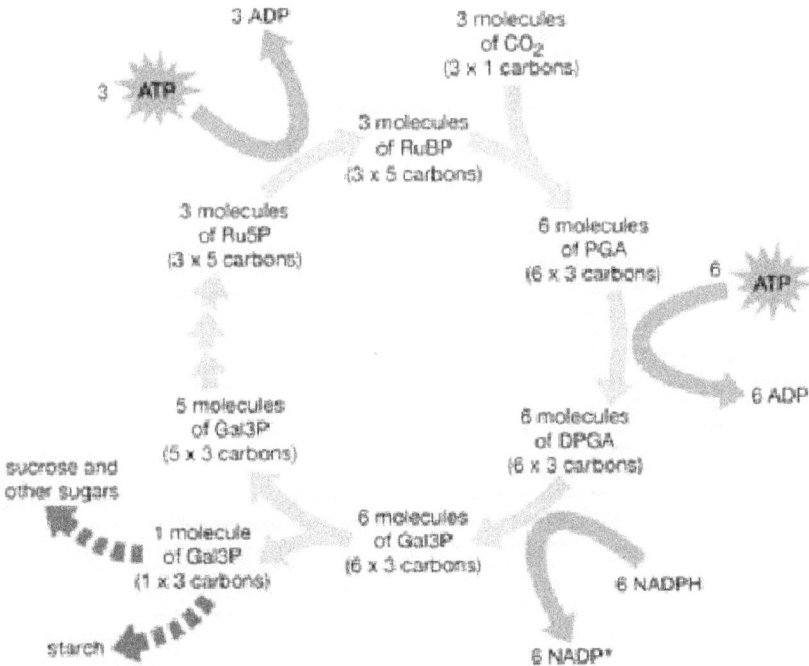

Pathway of carbon dioxide fixation and reduction in photosynthesis, the Calvin cycle. The diagram represents one complete turn of the cycle, with the net production of one molecule of glyceraldehyde-3-phosphate (Gal3P). This three-carbon sugar phosphate usually is converted to either sucrose or starch.

Metabolic cycles in general—the extraction by organisms of useful energy and food molecules from environmental source material—can be described in terms of oxidation-reduction reactions. In the case of oxygen respiration, oxygen molecules from the air accept electrons ultimately from glucose or amino acids. The oxygen, which has a great affinity for electrons, is called an electron acceptor, whereas the glucose, or other sugar or organic molecules, is an electron donor. Animal respiration is the prototype of oxidation-reduction reactions, but certainly not all oxidation-reduction reactions (or "redox reactions," as they are often called) involve oxygen. Many other inorganic compounds are respired, or "breathed," at the cell level. Biological electron acceptors besides oxygen include nitrate, nitrite, sulfate, carbonate, elemental sulfur, and methanol. Biological electron donors (other than sugar and amino acids) include hydrogen, nitrogen compounds (as ammonia, nitrite), sulfide, and methane. For acceptor-donor transformations to be available to chemoautotrophs and heterotrophs over sustained periods of time, ecological cycles are required. For geologically short periods of time, organisms may live off a finite supply of material; however, for any long-term continuance of life, a dynamic cycling of matter involving complementary types of organisms must prevail. If life exists on other planets, the requisite elements and liquid water must cycle. A search for such transformations provides one method of detecting extraterrestrial life.

In addition to energy, all forms of life require carbon sources. Autotrophic organisms (chemosynthetic and photosynthetic bacteria, algae, and plants) derive this essential element from carbon dioxide. Heterotrophs use preformed organic compounds as their source of carbon. Among autotrophs many types of cells do not depend on light to generate ATP; those that do without light are the chemoautotrophic bacteria, including the methanogens, ammonia oxidizers, sulfide oxidizers, hydrogen oxidizers, and a few obscure others. Indeed, at least five metabolic pathways entirely different from each other have evolved to use carbon dioxide gas. One is the oxygenic pathway described above, which is used by plants, algae, and cyanobacteria: the Calvin-Benson dark reactions. Other, more obscure pathways include phosphoenolpyruvate (PEP), succinate, and methanogen pathways. They all need to bring energy-poor carbon dioxide into the energy-rich carbon-hydrogen compound metabolism of organisms. All life on Earth depends on these autotrophic reactions that begin with carbon dioxide or its equivalent. Equivalents as carbon sources in autotrophic metabolism include the carbonate ion, bicarbonate ion, and carbon monoxide. As usual, with respect to metabolic variation and virtuosity, the bacterial repertoire is vastly more diverse than that of eukaryotes—that is, plants, animals, and other organisms composed of cells with nuclei. In general, nucleated organisms, eukaryotes, are either photolithoautotrophs (i.e., algae and plants) that derive energy from light or minerals or chemo-organoheterotrophs (animals, fungi, and most protists) that derive energy and carbon from preformed organic compounds (food).

ATP

All ATP biological electron-transfer reactions lead to the net production of ATP molecules. Two of the three phosphates (PO4) of this molecule are held by energy-rich bonds sufficiently stable to survive for long periods of time in the cell but not so strong that the cell cannot tap these bonds for energy when needed. ATP and similar molecules (such as guanosine triphosphate [GTP]) have a five-carbon sugar and three phosphates. As far as is known, such molecules are the general and unique energy currency of living systems on Earth.

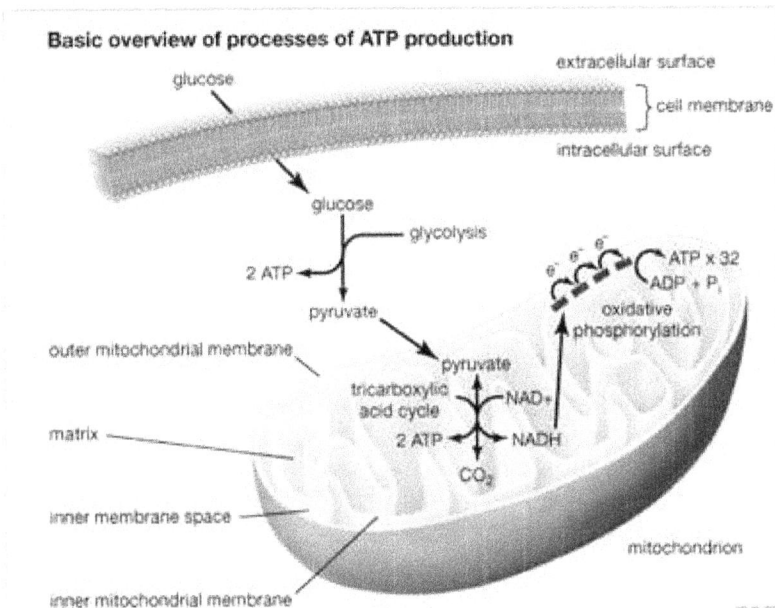

Basic overview of processes of ATP production

glucose
extracellular surface
cell membrane
intracellular surface
glucose
glycolysis
2 ATP
pyruvate
outer mitochondrial membrane
ATP x 32
ADP + P₁
oxidative phosphorylation
pyruvate
tricarboxylic acid cycle
NAD+
matrix
2 ATP
NADH
CO₂
inner membrane space
inner mitochondrial membrane
mitochondrion

The three processes of ATP production include glycolysis, the tricarboxylic acid cycle, and oxidative phosphorylation. In eukaryotic cells the latter two processes occur within mitochondria. Electrons that are passed through the electron transport chain ultimately generate free energy capable of driving the phosphorylation of ADP.

No metabolic process occurs in a single step. The ordinary six-carbon sugar, glucose, does not oxidize to carbon dioxide and water in living cells in the same way that glucose in air burns. Any release of energy by burning would be too sudden and too concentrated in a small volume to happen safely inside the tiny cell. Instead, glucose is broken down at ambient (i.e., relatively cool) temperatures by a series of successive and coordinated steps. Each step is mediated by a particular and specific enzyme. In most cells that metabolize glucose, the sugar first breaks down in a set of steps that occur in the absence of oxygen. The total number of such steps in plants, animals, fungi, and protists is about 11. Other organisms, primarily bacteria and obscure protists and fungi, are anaerobes: they do not utilize molecular oxygen in their metabolism. In anaerobes, glucose metabolism stops at compounds such as ethanol or lactic acid. Aerobic organisms, including all animals, carry the oxidation of glucose farther. They rapidly use anaerobic glucose breakdown products such as lactic acid, ethanol, or acetate with Krebs-cycle intermediates in the mitochondria. Aerobic oxidation of glucose requires an additional 60 enzyme-catalyzed steps. The anaerobic breakdown of glucose uses enzymes suspended freely in solution in the cells. The aerobic steps occur on enzymes localized in mitochondria, the "power packs" of cells where oxygen gas is used to make the energy compound ATP. The complete aerobic breakdown of sugar to carbon dioxide and water is about 10 times more efficient than the anaerobic in that 10 times as many ATP molecules are produced.

Energy made available to cells in the form of ATP is used in a variety of ways—for example, for motility. When an amoeba extends pseudopods or when a person walks, ATP molecules are tapped for their energy-rich phosphate bonds. ATP molecules are used for the synthesis of proteins that all cells require in their growth and division, amino acids, and five-carbon sugars of nucleic acids. Each synthetic process is controlled and enzymatically mediated. Each starts from an organic building-block compound available to the cell as food.

The amino acid L-leucine, for example, is produced from pyruvic acid, which is itself the product of the anaerobic breakdown of glucose. Synthesis of L-leucine from pyruvic acid involves eight enzyme-mediated steps with an addition of acetic acid and water.

These exquisitely interlocked and controlled metabolic steps are not performed in a diffuse manner helter-skelter in the cell. Rather, a marvelously architectured cellular interior displays specialized regions visible at the electron-microscopic level. Particular chemical reactions are performed in association with specific structures. In aerobic eukaryotes the mitochondrion with its intricate cristate membrane (the folds in the membrane are called cristae) provides the site of pyruvate, acetate, and lactate metabolism. These molecules are transformed and passed on from one enzyme to another as through a conveyor belt in a factory. Similarly, in those eukaryotes capable of oxygenic photosynthesis (algae, plants), photosynthesis occurs only in an organelle (a cell part) called a chloroplast. Chlorophyll, carotenoids, and other pigments that absorb visible light, as well as the detailed enzymatic apparatus for the photosynthetic process, reside there.

Chloroplasts and mitochondria contain DNA. Moreover, this DNA has a sequence distribution that differs entirely from that of the nucleus and greatly resembles that of free-living photosynthetic and oxygen-respiring bacteria. The best explanation for these facts is that the ancestors to the oxygen-releasing chloroplasts and oxygen-respiring mitochondria were once free-living bacteria.

Diversity
Prokaryotes and Eukaryotes
All life is composed of cells of one of two types: prokaryotes (those that lack a nucleus) or eukaryotes (those with a nucleus). Even in one-celled organisms this distinction is very clear.

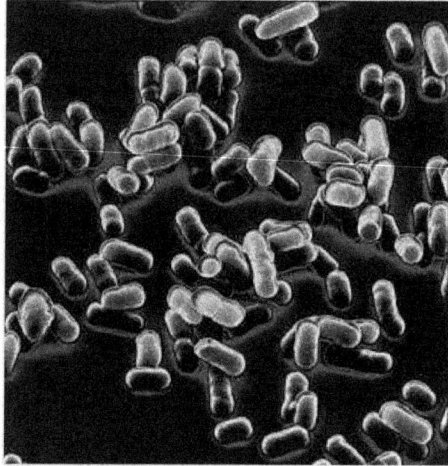

Scanning electron micrograph of
Yersinia pestis, the bacterium responsible for plague.

All bacteria are prokaryotic, even though many, probably most, are multicelled in nature. The only other single-celled organisms that exist are fungi (one-celled fungi are called yeasts). All nucleated organisms (cells with nuclei and chromosomes in their cells) that are not animals, fungi, or plants are Protista. This huge group includes the unicellular or few-celled protists and their multicellular descendants. The large kingdom of Protista has 250,000 estimated species alive today. Some are very large, such as red algae and the kelp Macrocystis. One-celled protists include the familiar amoebas, paramecia, and euglenas as well as 50,000 less-familiar types. Scientifically speaking, no such thing as a one-celled animal exists. All animals and plants are by definition multicellular, since they all develop from multicellular embryos. Accordingly, all "protozoans" are now classified as single-celled Protista, not animals. Nor are there any one-celled plants. Organisms formerly called one-celled plants are algae and, as such, are now classified with Protista as well. If a mature organism is determined to be one-celled, then it must be either a bacterium (prokaryotic) or a fungus or protist (eukaryotic). All animals and plants develop from embryos that by definition combine two complementary sets of chromosomes (i.e., they are diploids at some stage in their development). They are all multicellular eukaryotes. But though there are no one-celled plants or animals, there are indeed myriad many-celled protists. Multicellularity evolved not only in the ancestors to the plants and the animals but also in the bacteria, the protists, and the fungi.

All eukaryotic cells undergo some form of mitosis, a sequence of cell division events that occurs after chromosomal DNA protein replication. Mitosis ensures that chromosomal DNA and protein are equally distributed to the offspring cells. Mitosis is the most distinctive activity of eukaryotic cells, which have nucleoprotein chromosomes in their nuclei and a membrane that separates the nucleus from the cytoplasm. In mitosis, mitochondria, which are usually present in the cytoplasm as well as in the chloroplasts of algae and plants, are smoothly distributed with the chromosomes to offspring cells. The Golgi apparatus and endoplasmic reticulum (ER), an intricately convoluted structure, serve to anchor many cytoplasmic enzymes excluded from mitochondria or chloroplasts. They also divide and are distributed in mitosis.

Stages of mitosis. A. Prophase. Replicated chromosomes, consisting of two daughter strands (chromatids) attached by a centromere, coil and contract. Two pairs of specialized organelles (centrioles) begin to move apart, forming a bridge of hollow protein cylinders known as microtubules (spindle fibres) between them. Microtubules also extend in a radial array (aster) from the centrioles to the poles of the cell. B. Late prophase. As the centrioles move apart, the nuclear membrane breaks down and microtubules extend from each centromere to opposite sides or poles of the cell. C. Metaphase. The centromeres align in a plane midway between the poles known as the equator, or metaphase plate. During late metaphase, each centromere divides into two, freeing sister chromatids from each other. D. Anaphase. Sister chromatids are drawn to opposite ends as centromeric microtubules shorten and polar microtubules lengthen, causing the poles to move farther apart. E. Telophase. Chromosomes uncoil, microtubules disappear, and the nuclear membrane re-forms around each set of daughter chromosomes. The cytoplasm begins to pinch in to create two daughter cells.

Nuclei, chromosomes, mitochondria, chloroplasts, ER, and nuclear membranes are all absent in prokaryotes. Prokaryotic cells, which include all the cyanobacteria (formerly called blue-green algae), are bacteria in every way. Division is nonmitotic in all prokaryotes. Bacteria lack nucleoprotein and a nuclear membrane, and, when chromosome stain is applied, only fuzz or nothing is seen. Whereas all eukaryotic cells have more than one chromosome and sometimes over a thousand, the genes of prokaryotic cells are organized into a single "chromoneme" or "genophore." (The term bacterial chromosome, while still in use, is, technically speaking, inaccurate.) The genes may or may not be concentrated enough to be seen, but in any case bacterial DNA floats freely in the cytoplasm. Prokaryote cell organization is less complex than that of eukaryotes. The basic question of the evolution of prokaryotes into eukaryotes—often rated the second major evolutionary mystery, after the origin of life—is thought to involve a complex series of partnerships in which distinct strains of bacteria entered each others' bodies, merged symbiotically, and traded genes.

Multicellularity

Since multicellularity evolved independently in every major group of microorganisms, the blurred distinction between single-celled and many-celled organisms has become obsolete. The protists are divisible into about 35 unambiguous groups called phyla. They provide many examples of biological principles—including the prevalence of independent trends toward multicellularity. One illustration involves cellular slime molds. These heterotrophs undergo an extraordinary sequence of events during their life history. The story begins with single cells, indistinguishable from common amoebas. When starved, they begin to swarm. Soon they combine into a slimy mass of many nucleated amoeba cells called a pseudoplasmodium. The pseudoplasmodium in turn forms a sluglike multicellular creature resembling a mollusk that has escaped from its shell. This slug, which is entirely multicellular, migrates and then stops and develops into a stalk structure called a sorocarp that bears amoeba cysts on top. The cysts were called "spores." Some have cellulose cell walls similar to those of plants. The cysts, which are encased amoebas (just like other amoeba cysts), germinate in turn—when water and food again become abundant—into new amoebas. The released amoebas extend their pseudopods, and, as individuals again, they migrate to feed. The life history repeats with swarms of migrating amoebas, slugs, stalks, and finally clusters of amoeba cysts on top as wet, food-rich conditions are followed by dryness and scarcity.

Biology is replete with life histories of comparable or even greater complexity. In protist life histories—by far the most diverse, exotic, and unique—one can search for ancestral modes of life, including missing links between the prokaryotes and eukaryotes. The one-celled swimming stage is called a sperm, whose imperative it is to find another one-celled partner, the ovum. Like all animals, humans develop from a single fertilized egg with its complement of two sets of genetic material. These diploid fertilized egg cells then divide to form many presumably identical cells. The early embryology of all animals goes through stages that have 2, 4, 8, 16, and so on cells. Genetic information is theoretically identical in each cell. But then how does it ever happen that, as they mature, the cells become permanently specialized to form hair, bone, liver, blood, or nerve cells? How does any given cell "know" what sort of specialized cell it must become, since all cells seem to contain identical nucleic acids? Despite a century of work on this process (called differentiation) and the discovery of many facts about embryos, this basic problem still remains unsolved in animal biology.

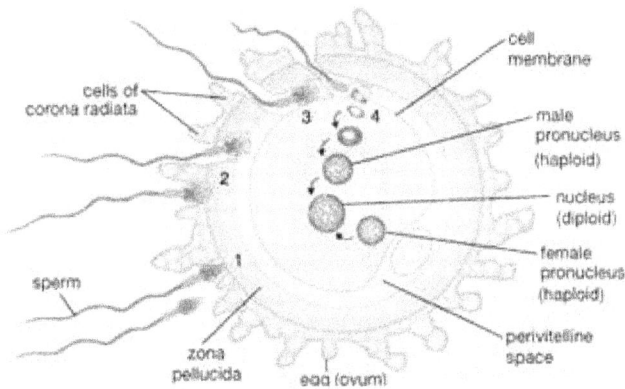

Fertilization of a human egg. (1) The sperm release enzymes that help disperse the corona radiata and bind to the zona pellucida. (2) The outer sperm head layer is sloughed off, exposing enzymes that digest a path through the zona pellucida. (3) The sperm fuses with the egg cell membrane, causing the zona pellucida to become impenetrable to other sperm. (4) The tail separates from the sperm head, and the male pronucleus enlarges and travels to the female pronucleus in the center of the cell. Chromosomes merge to form a fertilized egg.

Classification and Microbiota

Prior to the recognition of microbial life, the living world was too easily divided into animals that moved in pursuit of food and plants that produced food from sunlight. The futility of this simplistic classification scheme has been underscored by entire fields of science. Many bacteria both swim (like animals) and photosynthesize (like plants), yet they are best considered neither. Many algae (e.g., euglenids, dinomastigotes, chlorophytes) also swim and photosynthesize simultaneously. Molecular biological measurements of the DNA that codes for components of the ribosomes (organelles that are universally distributed) consistently find fungi to be extremely different from plants. Indeed, fungi genetically resemble animals more than plants.

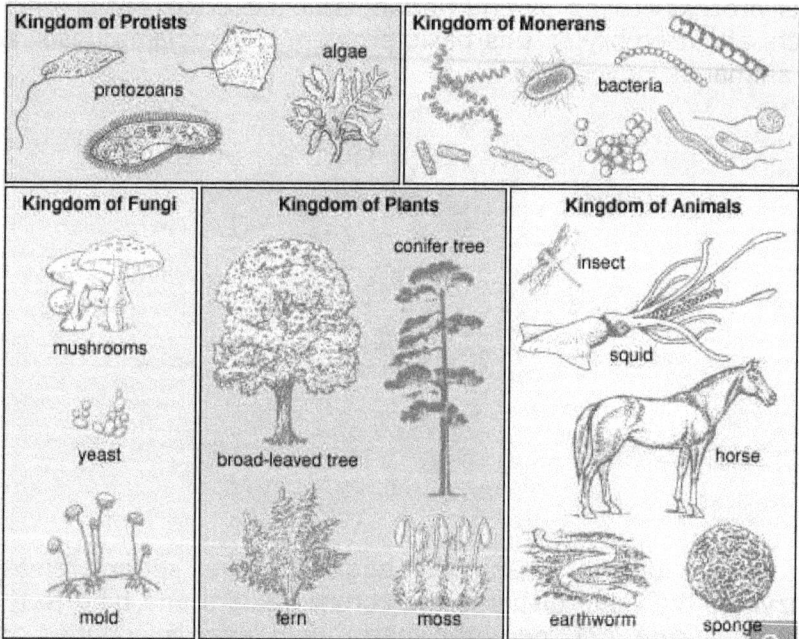

The Whittaker Five-Kingdom Classification of Life

Modern biology, following the lead of the German biologist Ernst Haeckel and the American biologists Herbert F. Copeland and Robert H. Whittaker, has now thoroughly abandoned the two-kingdom plant-versus-animal dichotomy. Haeckel proposed three kingdoms when he established "Protista" for microorganisms. Copeland classified the microorganisms into the Monerans (prokaryotes) and the Protoctista (which included fungi with the rest of the eukaryotic microorganisms). His four-kingdom scheme (Monera, Protoctista, Animalia, and Plantae) had the advantage of clearly separating microbes with nuclei (Protoctista) from those without (Monera: the prokaryotes—that is, the bacteria and archaea) and of distinguishing the two embryo-forming groups—plants and animals—from the rest of life. Whittaker, on ecological grounds, raised the fungi to kingdom status. The modified Whittaker five-kingdom classification system is perhaps the most comprehensible and biologically based way to unambiguously organize information about all groups of living beings. American microbiologist Carl Woese has offered still another classification scheme, in which all organisms are placed in either the Archaea (prokaryotes that include some salt lovers, acid lovers, and methane producers), the Bacteria (all other prokaryotes, including obligate anaerobic bacteria as well as photosynthetic and chemoautotrophic bacteria), or the Eukarya (all eukaryote forms of life). Woese's scheme is based on molecular biological criteria that focus on the RNA sequence of morphological factors to classify new or disputed organisms. Although Woese's three-domain system is very popular, a potential problem with it is that RNA, one characteristic among thousands, does not consistently correlate with many others.

Microbes (or microbiota) are simply all those organisms too small to be visualized without some sort of microscopy. Bacteria, the smaller fungi, and the smaller protists are undoubtedly microbes. Some scientists classify tiny animals, worms, and rotifers as microbes as well. Like weed, a plant not wanted in a garden, microbe is often a more useful term than one with a precise scientific meaning.

Sex
The world of microbes, in any case, is more vast, complex, diverse, and widespread than the visible ordinary world of plants and animals. For example, microbes have sexual lives that are different from those of the animal and plant kingdoms. In all organisms composed of prokaryotic cells, DNA that is not complexed with protein ("naked," or chromonemal, DNA) transfers from a source (such as a plasmid, a virus, a second cell, or even DNA molecules suspended in a solution) to a live prokaryotic cell. The recipient cell at the end of the sex act contains some quantity of its own DNA and integrates some

from the donor. All prokaryotes can reproduce in the absence of any sex act.

In eukaryotes the sexual act requires the opening of membranes and the fusion of entire cells or at least of cell nuclei. The contribution from genes to the recombinant offspring is approximately 50 percent from each parent. From two to a dozen or so genders (in some species of paramecia) are present in any given sexual group. Although any given sex act requires at least two individuals, mating tends to be by pairs. Gender is understood to be those traits that predispose any organism to enter the sex act with any other. In multigender species only two genders or mating types enter a sex act at any one time. The rule is that in multigender species a mating requires any gender other than one's own. Individual cells or multicellular organisms of complementary genders, in principle, produce fertile offspring. The universal rule is that no offspring result from matings of individuals of the same gender. In protists and fungi, uniparental reproduction (i.e., reproduction of a single parent) can occur in the absence of any sexual act, but two-parent sex may prevail seasonally or under other given environmental conditions in many inclusive taxa (such as families, classes, or phyla). Members of all species of the plant and animal kingdoms develop from embryos that form from a sexual act between the parents, and therefore two-parent (biparental) sex is the rule. Biparental sexuality of plants and animals has likely preceded its loss in all cases where a plant or animal species has reverted to uniparental reproduction, as in rotifers, whiptail lizards, and hundreds of plants that reproduce by runners rather than by seed. This suggestion is based on the fact that, at the cell level, aspects of meiosis (required for two-parent sex) continue to occur.

Ploidy, the concept of the number of complete sets of genes organized into chromosomes, is inapplicable to prokaryotes. Ploidy in protists, depending on species, varies so greatly and regularly that it is obvious that sexual cycles evolved in this diverse group of eukaryotes. Fungal cell nuclei are haploid (one set of chromosomes) or dikaryotic (two distinct nuclei from two different parents, each with one set of chromosomes sharing the same cell). Plant cell nuclei have two sets of chromosomes (diploid, in the sporophyte generation) or one (haploid, in the gametophyte generation). Animal cell nuclei, except in the gametes (sperm and egg), tend to have two sets of chromosomes (they are diploid).

Viruses

Viral and plasmid nucleic acids pass from cell to cell where the DNA or RNA perform their replication and coding functions efficiently. These "small replicons"—viruses and plasmids—are essentially strands of nucleic acid with a protein coat that depend entirely on the host cell for their continual existence. Pieces of the genetic material, virus-sized, pass from one cell into another cell of the same kind. Traveling small replicons of DNA produce genetic and permanently heritable changes in their new locations. Alternatively, part of the virus nucleic acid may be permanently bound to the nuclear DNA of the cell in which it resides. Viruses may be thought of as degenerate forms that are highly specialized in order to live in specific host cells of free-living organisms. Only cells are capable of performing the metabolic tasks that viruses and plasmids require. Viruses and plasmids must use the genetic transcription apparatus of cells. Bacterial viruses, or bacteriophages, may be extremely efficient in turning the luckless bacterium from a "factory" for the production of more bacteria into a factory for making more virus particles. It may take no more than 10 minutes for a bacterium infected by a single virus to produce 100 new virus particles, bursting forth from the victim bacterium by destroying it. Plasmids do not burst forth; rather, they benignly incorporate their DNA into that of their host cell.

Limits to Life

Organisms generated by the same fundamental biochemistry survive, grow, and reproduce in an extraordinarily wide range of conditions on Earth. For example, an alga called Cyanidium caldarium, a eukaryotic and photosynthetic organism, thrives in concentrated solutions of hot sulfuric acid and colours a damp landscape turquoise after a wet volcanic explosion. A swimming relative, Cyanophora paradoxa, survives in nearly these extremes. Certain less-colourful bacteria and fungi can live in extremely acidic environments (pH 0–2.5), such as that of Rio Tinto near Huelva in Spain. Bright blue-green cyanobacteria of many kinds can grow vigorously in extremely alkaline environments (pH 10–13).

Temperature and Desiccation

Most familiar organisms on Earth are of course sensitive to extreme temperature in their surroundings. Mammals and birds have evolved internal regulation of their temperatures. Humans cannot tolerate body temperatures below 30 °C (86 °F) or above 40 °C (104 °F). Cold-climate organisms have special insulating layers of fat and fur. Other organisms adjust to seasonal temperature drops by developing dormant propagules such as spores, eggs, or tuns, which are hardy desiccation- and radiation-resistant forms produced by microscopic animals called tardigrades, also known as "water bears." Dormancy is often accompanied by dehydration.

Most organisms are composed of an estimated 70–80 percent water. The availability of body water is a biological imperative. Certain halophilic bacteria live on water adsorbed on a single crystal of salt. Others such as the kangaroo rat (a mammal) and Tribolium (the flour beetle) imbibe no water at all in the liquid state. They rely entirely on metabolic water— that is, on water released from chemical bonds through the metabolism of food. A variety of plants, including Spanish moss, live without contact with groundwater. They extract water directly from the air, although they do require relatively high humidity. Desert plants and other plants in very dry environments, such as the two-leaved Welwitschia of the Namib Desert, have evolved extensive root systems that absorb subsurface water from a great volume of adjacent soil.

Life has been detected in the stratosphere and in the ocean's major depths. Mud-loving photosynthetic bacteria live in pools at Yellowstone National Park at temperatures above 90 °C (194 °F), whereas some unidentified dark-dwelling marine bacteria at marine hot vents have been recorded to grow at a superhot temperature of 113 °C (235 °F). (Because the water is under pressure, it is not above the boiling point.) Sulfate-reducing bacteria taken from the ocean grow and reproduce at 104 °C (219 °F) under high hydrostatic pressure. Many organisms employ organic or inorganic antifreezes to lower the freezing point of their internal liquids. Many kinds can live at several tens of degrees below 0 °C (32 °F). Some insects use dimethyl sulfoxide as an antifreeze. Other organisms live in briny pools in which dissolved salts lower the freezing point. San Juan Pond in Antarctica, for example, contains about one molecule of calcium chloride for every two water molecules. Not until −45 °C (−49 °F) does the pond freeze. A type of cryophilic (cold-loving) bacteria that lives there continues to metabolize down to at least −23 °C (−9 °F). Biological activity does not cease at the freezing point of water. In certain sea urchins, some microtubule proteins form the tubules of mitosis best at −2 °C (28 °F), and some enzymes are actually more active in ice than in water. Many bacteria are routinely frozen at −80 °C (−112 °F). They are thawed with no decrease in activity. Freezing temperatures alone cause no damage. Rather, frozen water removes tissue fluidity and leaves dangerous salt concentrations in its wake. The combination of expansion and contraction attendant to freezing and thawing harms membranes. Some arthropods can be severely dehydrated and then revived simply by the addition of water. Once dehydrated, these animals can be brought to any temperature from close to absolute zero (−273 °C, or −460 °F) to above the boiling point of water (100 °C, or 212 °F) without apparent damage. When encysted in response to dehydration, these arthropods at first glance are indistinguishable from a weathered grain of sand.

Cyanobacteria in a hot spring at
Yellowstone National Park, Wyoming.
Yellowstone National Park

Bacteria and fungal spores have been discovered near the base of the stratosphere by balloon searches. Organisms sought at much higher altitudes (up to 30,000 metres [100,000 feet]) have been detected; they are few in number and are all propagules. Birds have been observed to fly at maximum altitudes of 8,200 metres (27,000 feet), and on Mount Everest jumping spiders have been found at 6,700 metres (22,000 feet). At the opposite extreme, ciliates, pout fish, crabs, and clams have been recovered from ocean depths where pressures are hundreds of times those found at sea level. At these depths no light penetrates, and the organisms, some of which are quite large with bioluminescent organs that glow in the dark, feed on particles of organic matter raining down from the upper reaches of the oceans. Others sustain themselves by their chemoautotrophic bacterial associations.

Radiation and Nutrient Deprivation

The radiation environment of Earth has provoked evolutionary responses in many types of organisms. Some bacteria are readily killed by the small amount of solar ultraviolet light that filters through Earth's atmosphere at wavelengths near 300 nanometres. To the continuing annoyance of nuclear physicists, the bacterium Deinococcus radiodurans thrives in the cooling pool of nuclear reactors amid radioactivity levels lethal to mammals. Some life avoids radiation by shielding: algae and some desert plants live under a superficial coating of soil or rock that is more transparent to visible light than to ultraviolet light. Many produce protective epithelial coatings. Most telling is the fact that some microbes and animals have active methods of repairing damage produced by radiation. Some of these repair mechanisms work in the dark; others require visible light. Nucleic acids of all organisms absorb ultraviolet light very effectively at a wavelength near 260 nanometres, which accounts for their ultraviolet sensitivity. The upper limit to the amount of ionizing radiation (which includes gamma rays, X-rays, and electrons) that an organism can receive without being killed is approximately 1,000,000 roentgens. Such an extraordinarily high dose can be withstood only by Deinococcus. Mammals are killed by vastly lower doses, probably because so much more can go wrong in a large and complex animal. For the whole body of a human being, a dose of some 400 roentgens causes radiation sickness and death in half of those exposed to this level. A thermonuclear weapon dropped on a populated area may deliver, through direct radiation and fallout, doses of a few hundred roentgens or more to people within a radius of some tens of kilometres of the target. Much smaller doses produce a variety of diseases as well as deleterious mutations in the hereditary material, the DNA of the chromosomes. The effect of small doses of radiation is apparently cumulative. Until very recently no human beings had lived in environments with large fluxes of ionizing radiation.

Sizes of Organisms

The sizes of organisms on Earth vary greatly and are not always easy to estimate. On the large end, great stands of poplar trees entirely connected by common roots are really a single organism. A variety of influences place an upper limit to the size of organisms. One is the strength of biological materials. Sequoia redwood trees, some of which exceed 90 metres (300 feet), are apparently near the upper limit of height for an organism. The Italian astronomer Galileo calculated in 1638 that a tree taller than roughly 90 metres would buckle under its own weight when displaced slightly from the vertical (for example, by a breeze). Because of the buoyancy of water, large animals such as whales are not presented with such stability problems. Other size-related difficulties arise. The volume of tissues to be nourished increases as the cube of the characteristic length of the organism, but the surface of the gut, which absorbs the ingested food, increases only as the square of the length for a fixed morphology. As an organism's length is increased, a point of diminishing returns is ultimately reached where nutrition is irreversibly impeded in an animal.

New work on genome sequences, the total amount and quality of all of the genes that make up a live being, permits more accurate assessment of the material basis of the theoretically smallest and simplest extant free-living organisms. The complete DNA sequences of a few extremely small free-living organisms are now known—e.g., Mycoplasma genitalium with its 480 genes. All the molecules necessary for metabolism must be present. The smallest free-living cells include the pleuropneumonia-like organisms (PPLOs). Whereas an amoeba has a mass of 5×10^{-7} gram (2×10^{-8} ounce), a PPLO, which cannot be seen without a high-powered electron microscope, weighs 5×10^{-16} gram (2×10^{-15} ounce) and is only about 100 nanometres across. PPLOs grow very slowly. Other, even smaller organisms that grow even more slowly would be extremely difficult to detect. An organism the size of a PPLO that has room for only about a hundred enzymes depends entirely upon the animal tissue in which it lives. A much smaller organism would have room for many fewer enzymes. Its ability to accomplish the functions required for autopoiesis in living systems would be severely compromised. Were there, however, an environment in which all the necessary organic building blocks and such energy sources as ATP were provided "free," then there might be a functioning organism substantially smaller than a PPLO. The inside of cells provides just such an environment, which explains why infectious agents, such as prions, plasmids, and viruses, may be substantially smaller than PPLOs. But it must be emphasized that viruses and their kin are not, even in principle, autopoietic.

Metabolites and Water

The range of organic molecules that organisms, especially microbes, can metabolize is very wide and occasionally includes foods such as formaldehyde or petroleum that seem unlikely from a human point of view. Pseudomonas bacteria are capable of using almost any organic molecule as a source of carbon and energy, provided only that the molecule is at least slightly soluble in water. Microorganisms cannot metabolize plastics, not because of any fundamental chemical prohibitions but probably because plastics have not been part of the environment of microorganisms for very long. A lack of oxygen is thought of as extremely deleterious to life, but this view is anthropocentric. Many bacteria are facultative anaerobes that can take their oxygen or leave it. Many other bacteria and protists are obligate anaerobes that are actually poisoned by oxygen.

Water, which is crucial for life, is the major molecule in all organisms. Unless a massive mineral skeleton is present, the dry matter of most organisms is about one-half carbon by weight. This reflects the fact that all organic molecules are composed of carbon bound at least to hydrogen. Metabolism uses a wide variety of other chemical elements. Amino acids are made of nitrogen and sulfur in addition to carbon, hydrogen, and oxygen. Nucleic acids are made of phosphorus in addition to hydrogen, nitrogen, oxygen, and carbon. Sodium, potassium, and calcium are used to maintain electrolyte balance and to signal cells. Silicon is used as a structural material in the diatom shell, the radiolarian and heliozoan spicule, and the chrysophyte exoskeleton. Iron plays a fundamental role in the transport of molecular oxygen as part of the hemoglobin molecule. In some ascidians (sea squirts), however, vanadium replaces iron. Ascidian blood also contains unusually large amounts of niobium, titanium, chromium, manganese, molybdenum, and tungsten. The vanadium and niobium compounds in ascidian blood may be adaptations to low oxygen levels. Some bacteria use selenium, tellurium, or even arsenic as electron acceptors. Others produce the fully saturated gas hydrides of carbon, arsenic, phosphorus, or silicon as a metabolic waste. Still others form compounds of carbon with such halogens as chlorine or iodine. Not only the foregoing elements but also copper, zinc, cobalt, and possibly gallium, boron, and scandium perform particular functions in the enzymatic apparatus of particular cells. These elements, both the uncommon ones and those as common as phosphorus, are much more concentrated in living matter than in the environment where the living matter resides. This concentration suggests that such rare chemicals play unique functional roles that other, more abundant elements cannot serve.

Sensory Capabilities and Awareness

Although any given organism is severely limited in its range of behaviour patterns and sensory capabilities, life as a whole is remarkably sensitive to aspects of its local social and physical environment. A bird raised from the egg in the absence of other members of its species migrates when the season beckons, builds the proper nest, and engages in elaborate courtship rituals. Those birds that fail to perpetuate the behaviour pattern do not leave descendants. Such behavioural accuracy itself must have evolved. Rats that pass through mazes easily interbreed, as do rats that pass through with difficulty; eventually two populations with inherited characteristics called "maze-smart" and "maze-dumb" are produced. Fruit fly populations attracted to light can be separated from those that avoid light. Classical genetic-crossing experiments reveal that the two populations differ largely in a small number of genes for phototropism. Similar genetic determinants of behaviour exist in humans. For example, possession of a supernumerary Y-chromosome in males is strikingly correlated with aggressive tendencies.

Photosensitivity, Audiosensitivity, Thermosensitivity, Chemosensitivity, and Magnetosensitivity

Humans use only a limited region of the electromagnetic spectrum, the part called visible light, which extends from 400 to 700 nanometres in wavelength. While plants, algae, photosynthetic bacteria, and most animals are sensitive to this same range of wavelengths, many are sensitive to other wavelengths as well. Many plants present flower patterns visible only in the ultraviolet range at wavelengths below 400 nanometres, where pollinating insects are sensitive. Honeybees use polarized light—which the unaided human eye is unable to detect—for direction finding on partly cloudy days. The "pit" of such pit vipers as the rattlesnake is an infrared (heat) receptor that serves as a direction finder. These reptiles sense the thermal radiation emitted by mammals and birds, their warm-blooded prey. Humans are entirely insensitive to this thermal radiation.

That some animals such as dogs are sensitive to sounds that the human ear cannot detect is obvious to those who use dog whistles. Bats emit and detect sound waves at ultrahigh frequencies, in the vicinity of 100,000 cycles per second, about five times the highest frequency to which the human ear is sensitive. Bats have echolocated their prey by use of these sounds for millions of years before humans invented radar and sonar. The audio receptors of many moths that are prey to bats respond only to the frequencies emitted by the bats. When the bat sounds are heard, the moths take evasive action. Dolphins communicate via a very wide frequency range. They employ a "click" echolocator.

Some species of animals enjoy highly specialized and exotic organs for the detection or transmission of sound. Dolphins and whales use their blowholes rather than their mouths to utter their sounds.

Smell and taste, or some form of detection of specific chemical molecules, are universal. The ultimate in olfactory specialization may be male moths, whose feathery antennae are underlain by splayed microtubules, each of which is covered by a membrane at the distal end. They smell essentially nothing except the epoxide compound called disparlure, the chemical sex attractant discharged by the female. Only 40 molecules per second need impact on the antennae to produce a marked response. One female silkworm moth need release only 10^{-8} gram (4×10^{-10} ounce) of sex attractant per second in order to attract every male silkworm moth within a few kilometres.

Magnetotactic bacteria sense Earth's magnetic field. North Pole-seeking bacteria swim toward the sediment-water interface as they follow the magnetic lines of force. South Pole-seeking flagellated magnetotactic bacteria do the same in the Southern Hemisphere. Since those studied are microaerophiles—i.e., they require oxygen in lower than ambient concentrations—pole seekers tend to arrive at oxygen-depleted sediment adequate for their continued growth and reproduction. Ultrastructural studies reveal magnetosomes, tiny single-domain crystals of magnetite, an iron oxide mineral sensitive to magnetic fields, or greigite, an iron sulfide mineral, in their cells. The magnetosomes are aligned along the axis of the cell and serve to orient the sensitive bacteria. All the different kinds of magnetotactic bacteria bear magnetosomes in their cells. Whether magnetotaxis is causal in the orientation of homing pigeons, dancing bees on cloudy days, or other instinctively orienting animals is under investigation.

Besides the familiar senses of sight, hearing, smell, taste, and touch, organisms have a wide variety of other senses. People have inertial orientation systems and accelerometers in the cochlear canal of the ear. The water scorpion (Nepa) has a fathometer sensitive to hydrostatic pressure gradients. Many plants have chemically amplified gravity sensors made of modified chloroplasts. Some green algae use barium sulfate and calcium ion detection systems to sense gravity. Fireflies and squids communicate with their own kind by producing changing patterns of light on their bodies. The nocturnal African freshwater fish Gymnarchus niloticus operates a dipole electrostatic field generator and a sensor to detect the amplitude and frequency of disturbances in turbulent waters.

Sensing With Technology

From the foregoing sample of sensibilities, it is clear that a vast repertoire of sensors in living beings confers upon their possessors an awareness of the environment that differs from humanity's. Humans, however, have an enhanced ability to extend their sensory and intellectual capabilities through the use of instrumentation far beyond those with which they are born.

The sensory system of Earth is expanded by the capabilities of human machines. From those that detect ionizing radiation, wind velocities, the taste of wine, the concentration of salt in solution, a few photons of light in a dark corridor, or the blood temperature of an infant to those that record microearthquakes, a lying smile, or the heat of a furnace, the sensory systems of the biosphere—nonhuman, human, and human-mediated—have augmented over time. Indeed, just as seeing-eye dogs transmit visual information to their blind owners, the sensory system of life extends far beyond any given species of animals and its machines to the entire sensitive biota in this pulsating biosphere. Sensitivity to sound, chemicals, heat, light, mechanical movement, magnetism, and charged particles has been tallied by many a hardworking scientist. Whether entire categories of sensory information are missing from that list is not entirely clear. Great sensitivity to the environment abounds even in those smallest life-forms, the bacteria. Life has been sensing and responding to its environment since its inception more than 3.7 billion to 3.5 billion years ago. Moreover, it is not clear at what point in evolutionary history, or where precisely among organisms, consciousness comes in. Humans are conscious and self-conscious. But are protists that choose certain shapes and sizes of glass beads over others conscious of their decision making? Charles Darwin recognized selection among various male suitors by females as instrumental in the evolution of sexual species, including birds and insects. The extent to which consciousness and choice making are important in evolution remains a matter for debate.

Evolution and the History of Life on Earth
Heritability
The evidence is overwhelming that all life on Earth has evolved from common ancestors in an unbroken chain since its origin. Darwin's principle of evolution is summarized by the following facts. All life tends to increase: more organisms are conceived, born, hatched, germinated from seed, sprouted from spores, or produced by cell division (or other means) than can possibly survive. Each organism so produced varies, however little, in some measurable way from its relatives. In any given environment at any given time, those variants best suited to that environment will tend to leave more offspring than the others. Offspring resemble their ancestors. Variant organisms will leave offspring like themselves. Therefore, organisms will diverge from their ancestors with time. The term natural selection is shorthand for saying that all organisms do not survive to leave offspring with the same probability. Those alive today have been selected relative to similar ones that never survived or procreated. All organisms on Earth today are equally evolved since all share the same ancient original ancestors who faced myriad threats to their survival. All have persisted since roughly 3.7 billion to 3.5 billion years ago during the Archean Eon (4 billion to 2.5 billion years ago), products of the great evolutionary process with its identical molecular biological bases. Because the environment of Earth is so varied, the particular details of any organism's evolutionary history differ from those of another species in spite of chemical similarities.

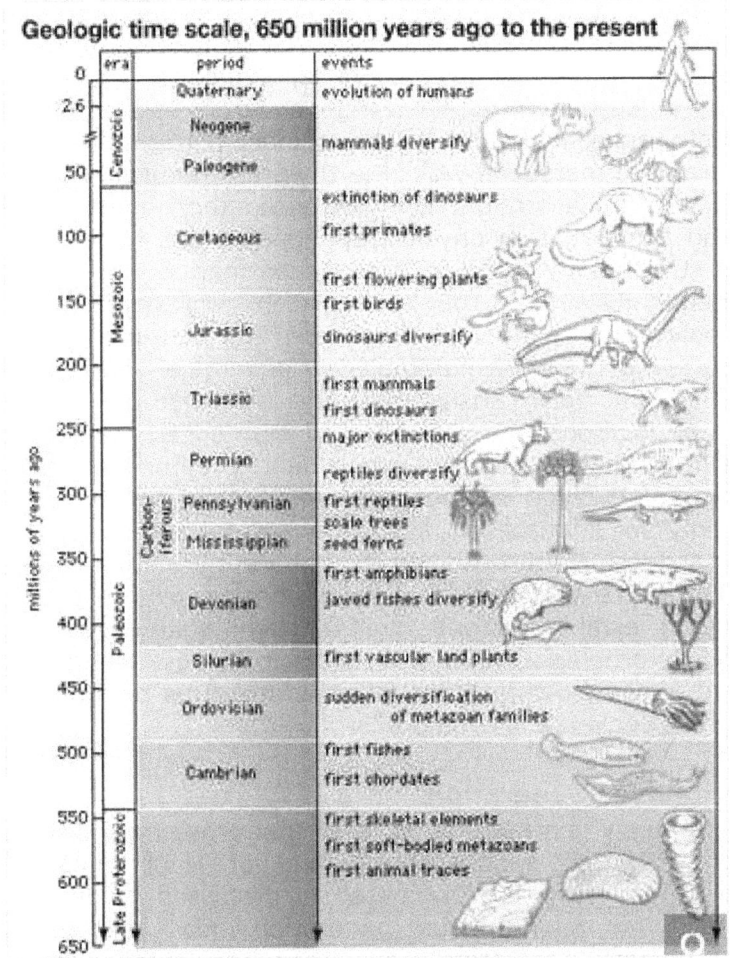

Geologic time scale, 650 million years ago to the present

The geologic time scale from 650 million years ago to the present, showing major evolutionary events.

Convergence

Everywhere the environment of Earth is heterogeneous. Mountains, oceans, and deserts suffer extremes of temperature, humidity, and water availability. All ecosystems contain diverse microenvironments: oxygen-depleted oceanic oozes, sulfide- or ammonia-rich soils, mineral outcrops with a high radioactivity content, or boiling organic-rich springs, for example. Besides these physical factors, the environment of any organism involves the other organisms in its surroundings. For each environmental condition, there is a corresponding ecological niche. The variety of ecological niches populated on Earth is quite remarkable. Even wet cracks in granite are replete with "rock eating" bacteria. Ecological niches in the history of life have been filled independently several times. For example, quite analogous to the ordinary placental mammalian wolf was the marsupial wolf, the thylacine (extinct since 1936) that lived in Australia; the two predatory mammals have striking similarities in physical appearance and behaviour. The same streamlined shape for high-speed marine motion evolved independently at least four times: in Stenopterygius and other Mesozoic reptiles; in tuna, which are fish; and in dolphins and seals, which are mammals. Convergent evolution in hydrodynamic form arises from the fact that only a narrow range of solutions to the problem of high-speed marine motion by large animals exists. The eye, a light receptor that makes an image, has evolved independently more than two dozen times not only in animals on Earth but in protists such as the dinomastigote Erythropsodinium. Apparently eyelike structures best solve the problem of visual recording. Where physics or chemistry establishes one most efficient solution to a given ecological problem, evolution in distinct lineages will often tend toward similar, nearly identical solutions. This phenomenon is known as convergent evolution.

Spontaneous Generation

Life ultimately is a material process that arose from a nonliving material system spontaneously—and at least once in the remote past. How life originated is discussed below. Yet no evidence for spontaneous generation now can be cited. Spontaneous generation, also called abiogenesis, the hypothetical process by which living organisms develop from nonliving matter, must be rejected. According to this theory, pieces of cheese and bread wrapped in rags and left in a dark corner were thought to produce mice, because after several weeks mice appeared in the rags. Many believed in spontaneous generation because it explained such occurrences as maggots swarming on decaying meat.

By the 18th century it had become obvious that plants and animals could not be produced by nonliving material. The origin of microorganisms such as yeast and bacteria, however, was not fully determined until French chemist Louis Pasteur proved in the 19th century that microorganisms reproduce, that all organisms come from preexisting organisms, and that all cells come from preexisting cells. Then what evidence is there for the earliest life on Earth?

Geologic Record

Past time on Earth, as inferred from the rock record, is divided into four immense periods of time called eons. These are the Hadean (4.6 billion to 4 billion years ago), the Archean (4 billion to 2.5 billion years ago), the Proterozoic (2.5 billion to 541 million years ago), and the Phanerozoic (541 million years ago to the present). For the Hadean Eon, the only record comes from meteorites and lunar rocks. No rocks of Hadean age survive on Earth. In the figure, eons are divided into eras, periods, and epochs. Such entries in the geologic time scale are often called "geologic time intervals."

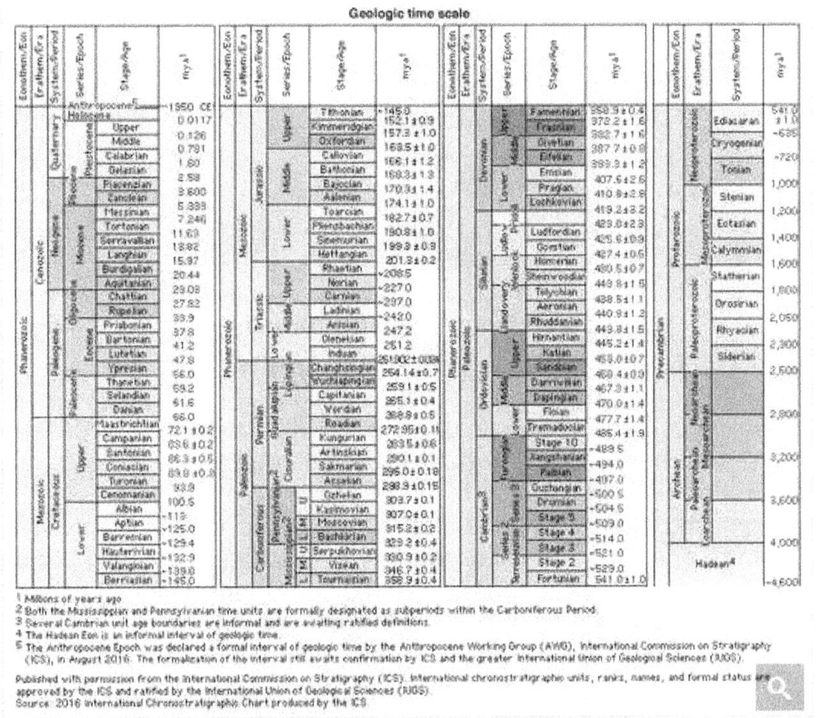

Geologic time scale

Geologic Time
The stratigraphic chart of geologic time.

Among the oldest known fossils are those found in the Fig Tree Chert from the Transvaal, dated over three billion years ago. These organisms have been identified as bacteria, including oxygenic photosynthetic bacteria (cyanobacteria)—i.e., prokaryotes rather than eukaryotes. Even prokaryotes, however, are exceedingly complicated organisms that grow and reproduce efficiently. Structures of communities of microorganisms, layered rocks called stromatolites, are found from more than three billion years ago. Since Earth is about 4.6 billion years old, these finds suggest that the origin of life must have occurred within a few hundred million years of that time.

Living stromatolites in Hamelin Pool
of Shark Bay, Western Australia.

Chemical analyses on organic matter extracted from the oldest sediments show what sorts of organic molecules are preserved in the rock record. Porphyrins have been identified in the oldest sediments, as have the isoprenoid derivatives pristane and phytane, breakdown products of cell lipids. Indications that these organic molecules dating from 3.1 billion to 2 billion years ago are of biological origin include the fact that their long-chain hydrocarbons show a preference for a straight-chain geometry. Chemical and physical processes alone tend to produce a much larger proportion of branched-chain and cyclic hydrocarbon molecular geometries than those found in ancient sediments. Nonbiological processes tend to form equal amounts of long-chain carbon compounds with odd and even numbers of carbon atoms. But products of undoubted biological origin, including the oldest sediments, show a distinct preference for odd numbers of carbon atoms per molecule. Another chemical sign of life is an enrichment in the carbon isotope C12, which is difficult to account for by nonbiological processes and which has been documented in some of the oldest sediments. This evidence suggests that bacterial photosynthesis or methanogenesis, processes that concentrate C12 preferentially to C13, were present in the early Archean Eon.

The Proterozoic Eon, once thought to be devoid of fossil evidence for life, is now known to be populated by overwhelming numbers of various kinds of bacteria and protist fossils—including acritarchs (spherical, robust unidentified fossils) and the entire range of Ediacaran fauna. The Ediacarans—large, enigmatic, and in some cases animal-like extinct life-forms—are probably related to extant protists. Almost 100 species are known from some 30 locations worldwide, primarily sandstone formations. Most Ediacarans, presumed to have languished in sandy seaside locales, probably depended on their internal microbial symbionts (photo- or chemoautotrophs) for nourishment. No evidence that they were animals exists. In addition to the Ediacarans, acritarchs, and other abundant microfossils, clear evidence for pre-Phanerozoic, or Precambrian, life includes the massive banded-iron formations (BIFs). Most BIFs date from 2.5 billion to 1.8 billion years ago. They are taken as indirect evidence for oxygen-producing, metal-depositing microscopic Proterozoic life. Investigations that use the electron microprobe (an instrument for visualizing structure and chemical composition simultaneously) and other micropaleontological techniques unfamiliar to classical geology have been employed to put together a much more complete picture of pre-Phanerozoic life.

Spriggina fossil from the Ediacaran
Period, found in the Ediacara Hills of Australia.

The earliest fossils are all of aquatic forms. Not until about two billion years ago are cyanobacterial filaments seen that colonized wet soil. By the dawn of the Phanerozoic Eon, life had insinuated itself between the Sun and Earth, both on land and in the waters of the world. For example, the major groups of marine animals such as mollusks and arthropods appeared for the first time about 541 million years ago at the base of the Cambrian Period of the Phanerozoic Eon. Plants and fungi appeared together in the exceptionally well-preserved Rhynie Chert of Scotland, dated about 408 million–360 million years ago in the Devonian Period. Solar energy was diverted to life's own uses. The biota contrived more and more ways of exploiting more and more environments. Many lineages became extinct. Others persisted and changed. The biosphere's height and depth increased, as did, by implication, the density of living matter. The proliferation and extinctions of a growing array of life-forms left indelible marks in the sedimentary rocks of the biosphere .

The Origin of Life
Hypotheses of origins

Perhaps the most fundamental and at the same time the least understood biological problem is the origin of life. It is central to many scientific and philosophical problems and to any consideration of extraterrestrial life. Most of the hypotheses of the origin of life will fall into one of four categories:

The origin of life is a result of a supernatural event—that is, one irretrievably beyond the descriptive powers of physics, chemistry, and other science.

Life, particularly simple forms, spontaneously and readily arises from nonliving matter in short periods of time, today as in the past.

Life is coeternal with matter and has no beginning; life arrived on Earth at the time of Earth's origin or shortly thereafter.

Life arose on the early Earth by a series of progressive chemical reactions. Such reactions may have been likely or may have required one or more highly improbable chemical events.

Hypothesis 1, the traditional contention of theology and some philosophy, is in its most general form not inconsistent with contemporary scientific knowledge, although scientific knowledge is inconsistent with a literal interpretation of the biblical accounts given in chapters 1 and 2 of Genesis and in other religious writings. Hypothesis 2 (not of course inconsistent with 1) was the prevailing opinion for centuries. A typical 17th-century view follows:

It was not until the Renaissance, with its burgeoning interest in anatomy, that such spontaneous generation of animals from putrefying matter was deemed impossible. During the mid-17th century the British physiologist William Harvey, in the course of his studies on the reproduction and development of the king's deer, discovered that every animal comes from an egg. An Italian biologist, Francesco Redi, established in the latter part of the 17th century that the maggots in meat came from flies' eggs, deposited on the meat. In the 18th century an Italian priest, Lazzaro Spallanzani, showed that fertilization of eggs by sperm was necessary for the reproduction of mammals. Yet the idea of spontaneous generation died hard. Even though it was clear that large animals developed from fertile eggs, there was still hope that smaller beings, microorganisms, spontaneously generated from debris. Many felt it was obvious that the ubiquitous microscopic creatures generated continually from inorganic matter.

Maggots were prevented from developing on meat by covering it with a flyproof screen. Yet grape juice could not be kept from fermenting by putting over it any netting whatever. Spontaneous generation was the subject of a great controversy between the famous French bacteriologists Louis Pasteur and Félix-Archimède Pouchet in the 1850s. Pasteur triumphantly showed that even the most minute creatures came from "germs" that floated downward in the air, but that they could be impeded from access to foodstuffs by suitable filtration. Pouchet argued, defensibly, that life must somehow arise from nonliving matter; if not, how had life come about in the first place?

Pasteur's experimental results were definitive: life does not spontaneously appear from nonliving matter. American historian James Strick reviewed the controversies of the late 19th century between evolutionists who supported the idea of "life from non-life" and their responses to Pasteur's religious view that only the Deity can make life. The microbiological certainty that life always comes from preexisting life in the form of cells inhibited many post-Pasteur scientists from discussions of the origin of life at all. Many were, and still are, reluctant to offend religious sentiment by probing this provocative subject. But the legitimate issues of life's origin and its relation to religious and scientific thought raised by Strick and other authors, such as the Australian Reg Morrison, persist today and will continue to engender debate.

Toward the end of the 19th century, hypothesis 3 gained currency. Swedish chemist Svante A. Arrhenius suggested that life on Earth arose from "panspermia," microscopic spores that wafted through space from planet to planet or solar system to solar system by radiation pressure. This idea, of course, avoids rather than solves the problem of the origin of life. It seems extremely unlikely that any live organism could be transported to Earth over interplanetary or, worse yet, interstellar distances without being killed by the combined effects of cold, desiccation in a vacuum, and radiation.

Although English naturalist Charles Darwin did not commit himself on the origin of life, others subscribed to hypothesis 4 more resolutely. The famous British biologist T.H. Huxley in his book Protoplasm: The Physical Basis of Life (1869) and the British physicist John Tyndall in his "Belfast Address" of 1874 both asserted that life could be generated from inorganic chemicals. However, they had extremely vague ideas about how this might be accomplished. The very phrase "organic molecule" implied, especially then, a class of chemicals uniquely of biological origin. Despite the fact that urea and other organic (carbon-hydrogen) molecules had been routinely produced from inorganic chemicals since 1828, the term organic meant "from life" to many scientists and still does. In the following discussion the word organic implies no necessary biological origin. The origin-of-life problem largely reduces to determination of an organic, nonbiological source of certain processes such as the identity maintained by metabolism, growth, and reproduction (i.e., autopoiesis).

Darwin's attitude was: "It is mere rubbish thinking at present of the origin of life; one might as well think of the origin of matter." The two problems are in fact curiously connected. Indeed, modern astrophysicists do think about the origin of matter. The evidence is convincing that thermonuclear reactions, either in stellar interiors or in supernova explosions, generate all the chemical elements of the periodic table more massive than hydrogen and helium. Supernova explosions and stellar winds then distribute the elements into the interstellar medium, from which subsequent generations of stars and planets form. These thermonuclear processes are frequent and well-documented. Some thermonuclear reactions are more probable than others. These facts lead to the idea that a certain cosmic distribution of the major elements occurs throughout the universe. Some atoms of biological interest, their relative numerical abundances in the universe as a whole, on Earth, and in living organisms are listed in the table. Even though elemental composition varies from star to star, from place to place on Earth, and from organism to organism, these comparisons are instructive: the composition of life is intermediate between the average composition of the universe and the average composition of Earth. Ninety-nine percent of the mass both of the universe and of life is made of six atoms: hydrogen (H), helium (He), carbon (C), nitrogen (N), oxygen (O), and neon (Ne). Might not life on Earth have arisen when Earth's chemical composition was closer to the average cosmic composition and before subsequent events changed Earth's gross chemical composition?

Relative abundances of the elements (percent)			
atom	universe	life (terrestrial vegetation)	Earth (crust)
hydrogen	87	16	3
helium	12	0*	0
carbon	0.03	21	0.1
nitrogen	0.008	3	0.0001
oxygen	0.06	59	49
neon	0.02	0	0
sodium	0.0001	0.01	0.7
magnesium	0.0003	0.04	8
aluminum	0.0002	0.001	2
silicon	0.003	0.1	14
sulfur	0.002	0.02	0.7
phosphorus	0.00003	0.03	0.07
potassium	0.000007	0.1	0.1
argon	0.0004	0	0
calcium	0.0001	0.1	2
iron	0.002	0.005	18

*0 percent here stands for any quantity less than 10^{-6} percent.

The Jovian planets (Jupiter, Saturn, Uranus, and Neptune) are much closer to cosmic composition than is Earth. They are largely gaseous, with atmospheres composed principally of hydrogen and helium. Methane, ammonia, neon, and water have been detected in smaller quantities. This circumstance very strongly suggests that the massive Jovian planets formed from material of typical cosmic composition. Because they are so far from the Sun, their upper atmospheres are very cold. Atoms in the upper atmospheres of the massive, cold Jovian planets cannot now escape from their gravitational fields, and escape was probably difficult even during planetary formation.

Photograph of Jupiter taken by Voyager 1 on February 1, 1979, at a range of 32.7 million km (20.3 million miles). Prominent are the planet's pastel-shaded cloud bands and Great Red Spot (lower centre).

Earth and the other planets of the inner solar system, however, are much less massive, and most have hotter upper atmospheres. Hydrogen and helium escape from Earth today; it may well have been possible for much heavier gases to have escaped during Earth's formation. Very early in Earth's history, there was a much larger abundance of hydrogen, which has subsequently been lost to space. Most likely the atoms carbon, nitrogen, and oxygen were present on the early Earth, not in the forms of CO2 (carbon dioxide), N2, and O2 as they are today but rather as their fully saturated hydrides: methane, ammonia, and water. The presence of large quantities of reduced (hydrogen-rich) minerals, such as uraninite and pyrite, that were exposed to the ancient atmosphere in sediments formed over two billion years ago implies that atmospheric conditions then were considerably less oxidizing than they are today.

In the 1920s British geneticist J.B.S. Haldane and Russian biochemist Aleksandr Oparin recognized that the nonbiological production of organic molecules in the present oxygen-rich atmosphere of Earth is highly unlikely but that, if Earth once had more hydrogen-rich conditions, the abiogenic production of organic molecules would have been much more likely. If large quantities of organic matter were somehow synthesized on early Earth, they would not necessarily have left much of a trace today. In the present atmosphere—with 21 percent of oxygen produced by cyanobacterial, algal, and plant photosynthesis—organic molecules would tend, over geological time, to be broken down and oxidized to carbon dioxide, nitrogen, and water. As Darwin recognized, the earliest organisms would have tended to consume any organic matter spontaneously produced prior to the origin of life.

The first experimental simulation of early Earth conditions was carried out in 1953 by a graduate student, Stanley L. Miller, under the guidance of his professor at the University of Chicago, chemist Harold C. Urey. A mixture of methane, ammonia, water vapour, and hydrogen was circulated through a liquid solution and continuously sparked by a corona discharge mounted higher in the apparatus. The discharge was thought to represent lightning flashes. After several days of exposure to sparking, the solution changed colour. Several amino and hydroxy acids, familiar chemicals in contemporary Earth life, were produced by this simple procedure. The experiment is simple enough that the amino acids can readily be detected by paper chromatography by high school students. Ultraviolet light or heat was substituted as an energy source in subsequent experiments. The initial abundances of gases were altered. In many other experiments like this, amino acids were formed in large quantities. On the early Earth much more energy was available in ultraviolet light than from lightning discharges. At long ultraviolet wavelengths, methane, ammonia, water, and hydrogen are all transparent, and much of the solar ultraviolet energy lies in this region of the spectrum. The gas hydrogen sulfide was suggested to be a likely compound relevant to ultraviolet absorption in Earth's early atmosphere. Amino acids were also produced by long-wavelength ultraviolet irradiation of a mixture of methane, ammonia, water, and hydrogen sulfide. At least some of these amino acid syntheses involved hydrogen cyanide and aldehydes (e.g., formaldehyde) as gaseous intermediates formed from the initial gases. That amino acids, particularly biologically abundant amino acids, are made readily under simulated early Earth conditions is quite remarkable. If oxygen is permitted in these kinds of experiments, no amino acids are formed. This has led to a consensus that hydrogen-rich (or at least oxygen-poor) conditions were necessary for natural organic syntheses prior to the appearance of life.

Under alkaline conditions, and in the presence of inorganic catalysts, formaldehyde spontaneously reacts to form a variety of sugars. The five-carbon sugars fundamental to the formation of nucleic acids, as well as six-carbon sugars such as glucose and fructose, are easily produced. These are common metabolites and structural building blocks in life today. Furthermore, the nucleotide bases and even the biological pigments called porphyrins have been produced in the laboratory under simulated early Earth conditions. Both the details of the experimental synthetic pathways and the question of stability of the small organic molecules produced are vigorously debated. Nevertheless, most, if not all, of the essential building blocks of proteins (amino acids), carbohydrates (sugars), and nucleic acids (nucleotide bases) - that is, the monomers—can be readily produced under conditions thought to have prevailed on Earth in the Archean Eon. The search for the first steps in the origin of life has been transformed from a religious/philosophical exercise to an experimental science.

Production of Polymers
The formation of polymers, long-chain molecules made of repeating units of monomers (the essential building blocks mentioned above), is a far more difficult experimental problem than the formation of monomers. Polymerization reactions tend to be dehydrations. A molecule of water is lost in the formation of a peptide from two amino acids or of a disaccharid sugar from two monomers. Dehydrating agents are used to initiate polymerization. The polymerization of amino acids to form long proteinlike molecules ("proteinoids") was accomplished through dry heating by American biochemist Sidney Fox and his colleagues. The polyamino acids that he formed are not random molecules unrelated to life. They have distinct catalytic activities. Long polymers of amino acids were also produced from hydrogen cyanide and anhydrous liquid ammonia by American chemist Clifford Matthews in simulations of the early upper atmosphere. Some evidence exists that ultraviolet irradiation induces combinations of nucleotide bases and sugars in the presence of phosphates or cyanides. Some condensing agents such as cyanamide are efficiently made under simulated primitive conditions. Despite the breakdown by water of molecular intermediates, condensing agents may quite

effectively induce polymerization, and polymers of amino acids, sugars, and nucleotides have all been made this way.

That adsorption of relevant small carbon compounds on clays or other minerals may have concentrated these intermediates was suggested by the British scientist John Desmond Bernal. Concentration of some kind is required to offset the tendency for water to break down polymers of biological significance. Phosphorus, which with deoxyribose sugar forms the "backbone" of DNA and is integrally involved in cell energy transformation and membrane formation, is preferentially incorporated into prebiological organic molecules. It is hard to explain how such a preference could have happened without the concentration of organic molecules.

The early ocean and lakes themselves may have been a dilute solution of organic molecules. If all the surface carbon on Earth were present as organic molecules, or if many known ultraviolet synthetic reactions that produce organic molecules were permitted to continue for a billion years with their products dissolved in the oceans, a 1 percent solution of organic molecules would result. Haldane suggested that the origin of life occurred in a "hot dilute soup." Concentration through either evaporation or freezing of pools, adsorption on clay interfaces, or the generation of colloidal enclosures called coacervates may have served to bring the organic molecules in question in contact with each other.

The essential building blocks for life (the monomers) were probably produced in relatively abundant concentrations, given conditions on the early Earth. Although relevant, this is more akin to the origin of food than to the origin of life. If life is defined as a self-maintaining, self-producing, and mutable molecular system that derives energy and supplies from the environment, then food is certainly required for life. Polynucleotides (polymers of RNA and DNA) can be produced in laboratory experiments from nucleotide phosphates in the presence of enzymes of biological origin (polymerases) and a preexisting "primer" nucleic acid molecule. If the primer is absent, polynucleotides are still formed, but they lack specific genetic information. Once such a polynucleotide forms, it can act as a primer for subsequent syntheses.

Even if such a molecular population could replicate polynucleotides, it would not be considered alive. The polynucleotides tend to hydrolyze (break down) in water. In the early 1980s American biochemist Thomas Cech and Canadian

American molecular biologist Sidney Altman discovered that certain RNA molecules have catalytic properties. They catalyze their own splicing, which suggests an early role for RNA in life or even in life's origins. Only the partnership of the two kinds of molecules (proteins and nucleic acids) segregated from the rest of the world by an oily membrane makes the growth process of life on Earth possible. The molecular apparatus ancillary to the operation of the genetic code—the rules that determine the linear order of amino acids in proteins from nucleotide base pairs in nucleic acids (i.e., the activating enzymes, transfer RNAs, messenger RNAs, ribosomes, and so on)—may be the product of a long evolutionary history among natural, thermodynamically favoured, gradient-reducing complex systems. These rules are produced according to instructions contained within the code. American biophysicist Harold J. Morowitz argued cogently that the origin of the genetic system, the code with its elaborate molecular apparatus, occurred inside cells only after the origin of life as a cyclic metabolic system. American theoretical biologist Jeffrey Wicken pointed out that replicating molecules, if they appeared first, would have had no impetus to develop a complex cellular package or associated protein machinery and that life thus probably arose as a metabolic system that was stabilized by the genetic code, which allowed life's second law-favoured process to continue ad infinitum.

Many separate and rather diverse instances of the origin of living cells may have occurred in the Archean Earth, but obviously only one prevailed. Interactions eventually eliminated all but our lineage. From the common composition, metabolism, chemical behaviour, and other properties of life, it seems clear that every organism on Earth today is a member of the same lineage.

The Earliest Living Systems

Most organic molecules made by living systems inside cells display the same optical activity: when exposed to a beam of plane-polarized light, they rotate the plane of the beam. Amino acids rotate light to the left, whereas sugars, called dextrorotatory, rotate it to the right. Organic molecules produced artificially lack optical activity because both "left-handed" and "right-handed" molecules are present in equal quantity. Molecules of the same optical activity can be assembled in complementary ways like the stacking of right-handed gloves. The same monomers can be used to produce longer chain molecules that are three-dimensional mirror images of each other; mixtures of monomers of different handedness cannot. Cumulative symmetry is responsible for optical activity. At the time of the origin of life, organic molecules, corresponding both to left- and right-handed forms, were no doubt formed as they are in laboratory simulation experiments today: both types were produced. But the first living systems must have employed one type of component, for the same reason that carpenters cannot use random mixtures of screws with left- and right-handed threads in the same project with the same tools. Whether left- or right-handed activity was adopted was probably a matter of chance, but, once a particular asymmetry was established, it maintained itself. Optical activity accordingly is likely to be a feature of life on any planet. The chances may be equal of finding a given organic molecule or its mirror image in extraterrestrial life-forms if, as Morowitz suspects, the incorporation of nitrogen into the first living system involved glutamine, the simplest of the required amino acid precursors with optical activity.

The first living cells probably resided in a molecular Garden of Eden, where the prebiological origin of food had guaranteed monomers that were available. The cells, the first single-celled organisms, would have increased rapidly. But such an increase was eventually limited by the supply of molecular building blocks. Those organisms with an ability to synthesize scarce monomers, say A, from more abundant ones, say B, would have persisted. The secondary source of supply, B, would in time also become depleted. Those organisms that could produce B from a third monomer, C, would have preferentially persisted. The American biochemist Norman H. Horowitz has proposed that the multienzyme catalyzed reaction chains of contemporary cells originally evolved in this way

*****

Astronomy

Astronomy, science that encompasses the study of all extraterrestrial objects and phenomena. Until the invention of the telescope and the discovery of the laws of motion and gravity in the 17th century, astronomy was primarily concerned with noting and predicting the positions of the Sun, Moon, and planets, originally for calendrical and astrological purposes and later for navigational uses and scientific interest. The catalog of objects now studied is much broader and includes, in order of increasing distance, the solar system, the stars that make up the Milky Way Galaxy, and other, more distant galaxies. With the advent of scientific space probes, Earth also has come to be studied as one of the planets, though its more-detailed investigation remains the domain of the Earth sciences.

Hubble Space Telescope
Hubble Space Telescope,
photographed by the space shuttle Discovery.

The Scope of Astronomy

Since the late 19th century astronomy has expanded to include astrophysics, the application of physical and chemical knowledge to an understanding of the nature of celestial objects and the physical processes that control their formation, evolution, and emission of radiation. In addition, the gases and dust particles around and between the stars have become the subjects of much research. Study of the nuclear reactions that provide the energy radiated by stars has shown how the diversity of atoms found in nature can be derived from a universe that, following the first few minutes of its existence, consisted only of hydrogen, helium, and a trace of lithium. Concerned with phenomena on the largest scale is cosmology, the study of the evolution of the universe. Astrophysics has transformed cosmology from a purely speculative activity to a modern science capable of predictions that can be tested.

Its great advances notwithstanding, astronomy is still subject to a major constraint: it is inherently an observational rather than an experimental science. Almost all measurements must be performed at great distances from the objects of interest, with no control over such quantities as their temperature, pressure, or chemical composition. There are a few exceptions to this limitation—namely, meteorites (most of which are from the asteroid belt, though some are from the Moon or Mars), rock and soil samples brought back from the Moon, samples of comet and asteroid dust returned by robotic spacecraft, and interplanetary dust particles collected in or above the stratosphere. These can be examined with laboratory techniques to provide information that cannot be obtained in any other way. In the future, space missions may return surface materials from Mars, or other objects, but much of astronomy appears otherwise confined to Earth-based observations augmented by observations from orbiting satellites and long-range space probes and supplemented by theory.

Nickel-Iron Meteorite
Nickel-iron meteorite, from Canyon Diablo, Arizona.

Determining Astronomical Distances

A central undertaking in astronomy is the determination of distances. Without a knowledge of astronomical distances, the size of an observed object in space would remain nothing more than an angular diameter and the brightness of a star could not be converted into its true radiated power, or luminosity. Astronomical distance measurement began with a knowledge of Earth's diameter, which provided a base for triangulation.

Within the inner solar system, some distances can now be better determined through the timing of radar reflections or, in the case of the Moon, through laser ranging. For the outer planets, triangulation is still used. Beyond the solar system, distances to the closest stars are determined through triangulation, in which the diameter of Earth's orbit serves as the baseline and shifts in stellar parallax are the measured quantities. Stellar distances are commonly expressed by astronomers in parsecs (pc), kiloparsecs, or megaparsecs. (1 pc = 3.086×10^{18} cm, or about 3.26 light-years [1.92×10^{13} miles].) Distances can be measured out to around a kiloparsec by trigonometric parallax. The accuracy of measurements made from Earth's surface is limited by atmospheric effects, but measurements made from the Hipparcos satellite in the 1990s extended the scale to stars as far as 650 parsecs, with an accuracy of about a thousandth of an arc second. The Gaia satellite is expected to measure stars as far away as 10 kiloparsecs to an accuracy of 20 percent. Less-direct measurements must be used for more-distant stars and for galaxies.

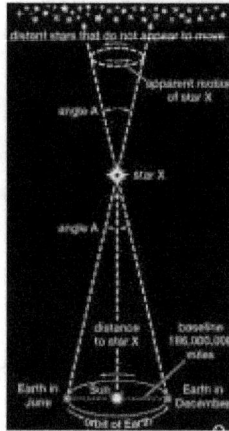

Stellar Distances

Calculating Stellar Distances

Two general methods for determining galactic distances are described here. In the first, a clearly identifiable type of star is used as a reference standard because its luminosity has been well determined. This requires observation of such stars that are close enough to Earth that their distances and luminosities have been reliably measured. Such a star is termed a "standard candle." Examples are Cepheid variables, whose brightness varies periodically in well-documented ways, and certain types of supernova explosions that have enormous brilliance and can thus be seen out to very great distances. Once the luminosities of such nearer standard candles have been calibrated, the distance to a farther standard candle can be calculated from its calibrated luminosity and its actual measured intensity. (The measured intensity [I] is related to the luminosity [L] and distance [d] by the formula $I = L/4\pi d2$.) A standard candle can be identified by means of its spectrum or the pattern of regular variations in brightness. (Corrections may have to be made for the absorption of starlight by interstellar gas and dust over great distances.) This method forms the basis of measurements of distances to the closest galaxies.

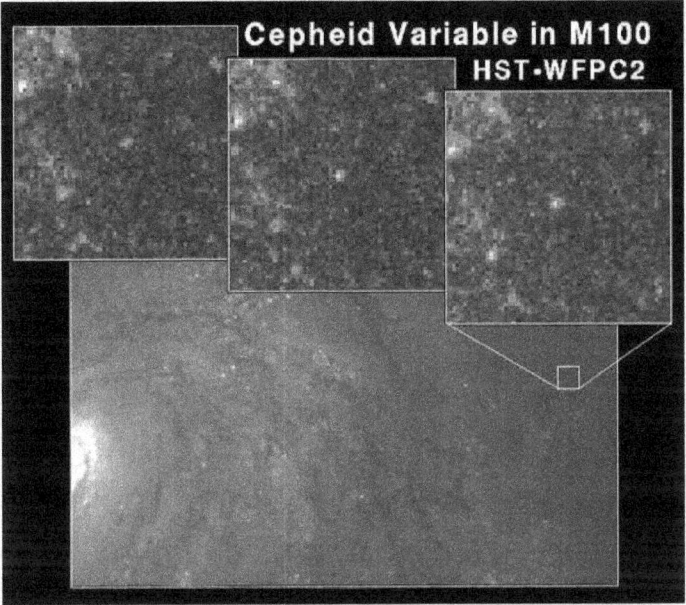

A region of the spiral galaxy M100 (bottom), with three frames (top) showing a Cepheid variable increasing in brightness. These images were taken with the Wide Field Planetary Camera 2 (WFPC2) on board the Hubble Space Telescope (HST).

The second method for galactic distance measurements makes use of the observation that the distances to galaxies generally correlate with the speeds with which those galaxies are receding from Earth (as determined from the Doppler shift in the wavelengths of their emitted light). This correlation is expressed in the Hubble law: velocity = H × distance, in which H denotes Hubble's constant, which must be determined from observations of the rate at which the galaxies are receding. There is widespread agreement that H lies between 67 and 73 kilometres per second per megaparsec (km/sec/Mpc). H has been used to determine distances to remote galaxies in which standard candles have not been found. (For additional discussion of the recession of galaxies, the Hubble law, and galactic distance determination)

The Doppler shift

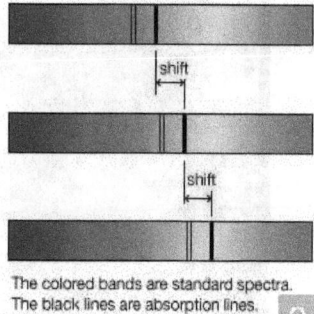

The amount of shift depends on the velocity of the object in relationship to the observer: the greater the velocity, the greater the shift.

Absorption lines from an approaching object shift toward the violet (shorter wavelength).

Absorption lines from the sun are used for comparison.

Absorption lines from a receding object shift toward the red (longer wavelength).

The colored bands are standard spectra. The black lines are absorption lines.

Doppler Shift.

Study of the Solar System

The solar system took shape 4.57 billion years ago, when it condensed within a large cloud of gas and dust. Gravitational attraction holds the planets in their elliptical orbits around the Sun. In addition to Earth, five major planets (Mercury, Venus, Mars, Jupiter, and Saturn) have been known from ancient times. Since then only two more have been discovered: Uranus by accident in 1781 and Neptune in 1846 after a deliberate search following a theoretical prediction based on observed irregularities in the orbit of Uranus. Pluto, discovered in 1930 after a search for a planet predicted to lie beyond Neptune, was considered a major planet until 2006, when it was redesignated a dwarf planet by the International Astronomical Union.

Jupiter, the fifth planet from the Sun and largest planet in the solar system. The Great Red Spot is visible in the lower left. This image is based on observations made by the Voyager 1 spacecraft in 1979.

The average Earth-Sun distance, which originally defined the astronomical unit (AU), provides a convenient measure for distances within the solar system. The astronomical unit was originally defined by observations of the mean radius of Earth's orbit but is now defined as 149,597,870.7 km (about 93 million miles). Mercury, at 0.4 AU, is the closest planet to the Sun, while Neptune, at 30.1 AU, is the farthest. Pluto's orbit, with a mean radius of 39.5 AU, is sufficiently eccentric that at times it is closer to the Sun than is Neptune. The planes of the planetary orbits are all within a few degrees of the ecliptic, the plane that contains Earth's orbit around the Sun. As viewed from far above Earth's North Pole, all planets move in the same (counterclockwise) direction in their orbits.

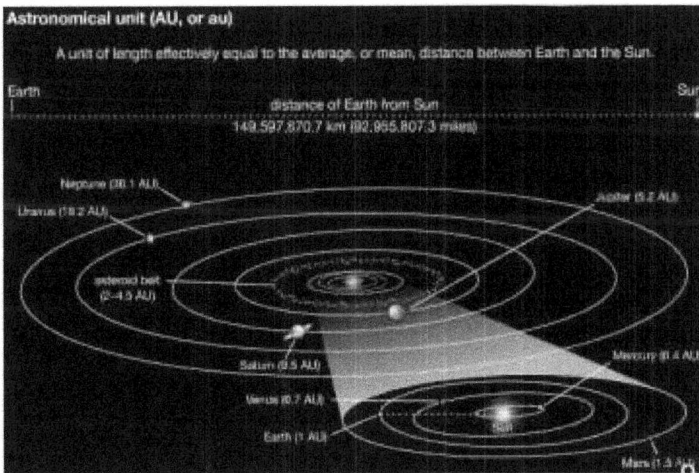

Scale of the Solar System

Distances in the solar system as measured in astronomical units (AU).

Most of the mass of the solar system is concentrated in the Sun, with its 1.99×10^{33} grams. Together, all of the planets amount to 2.7×10^{30} grams (i.e., about one-thousandth of the Sun's mass), and Jupiter alone accounts for 71 percent of this amount. The solar system also contains five known objects of intermediate size classified as dwarf planets and a very large number of much smaller objects collectively called small bodies. The small bodies, roughly in order of decreasing size, are the asteroids, or minor planets; comets, including Kuiper belt, Centaur, and Oort cloud objects; meteoroids; and interplanetary dust particles. Because of their starlike appearance when discovered, the largest of these bodies were termed asteroids, and that name is widely used, but, now that the rocky nature of these bodies is understood, their more descriptive name is minor planets.

The eight planets of the solar system and Pluto, in a montage of images scaled to show the approximate sizes of the bodies relative to one another. Outward from the Sun, which is represented to scale by the yellow segment at the extreme left, are the four rocky terrestrial planets (Mercury, Venus, Earth, and Mars), the four hydrogen-rich giant planets (Jupiter, Saturn, Uranus, and Neptune), and icy, comparatively tiny Pluto.

The eight planets of the solar system and Pluto, in a montage of images scaled to show the approximate sizes of the bodies relative to one another. Outward from the Sun, which is represented to scale by the yellow segment at the extreme left, are the four rocky terrestrial planets (Mercury, Venus, Earth, and Mars), the four hydrogen-rich giant planets (Jupiter, Saturn, Uranus, and Neptune), and icy, comparatively tiny Pluto.

The four inner, terrestrial planets—Mercury, Venus, Earth, and Mars—along with the Moon have average densities in the range of 3.9–5.5 grams per cubic cm, setting them apart from the four outer, giant planets—Jupiter, Saturn, Uranus, and Neptune—whose densities are all close to 1 gram per cubic cm, the density of water. The compositions of these two groups of planets must therefore be significantly different. This dissimilarity is thought to be attributable to conditions that prevailed during the early development of the solar system. Planetary temperatures now range from around 170 °C (330 °F, 440 K) on Mercury's surface through the typical 15 °C (60 °F, 290 K) on Earth to −135 °C (−210 °F, 140 K) on Jupiter near its cloud tops and down to −210 °C (−350 °F, 60 K) near Neptune's cloud tops. These are average temperatures; large variations exist between dayside and nightside for planets closest to the Sun, except for Venus with its thick atmosphere.

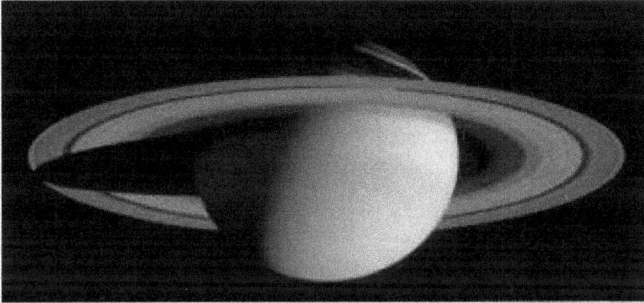

Saturn and its spectacular rings, in a natural-colour composite of 126 images taken by the Cassini spacecraft on October 6, 2004. The view is directed toward Saturn's southern hemisphere, which is tipped toward the Sun. Shadows cast by the rings are visible against the bluish northern hemisphere, while the planet's shadow is projected on the rings to the left.

The surfaces of the terrestrial planets and many satellites show extensive cratering, produced by high-speed impacts (see meteorite crater). On Earth, with its large quantities of water and an active atmosphere, many of these cosmic footprints have eroded, but remnants of very large craters can be seen in aerial and spacecraft photographs of the terrestrial surface. On Mercury, Mars, and the Moon, the absence of water and any significant atmosphere has left the craters unchanged for billions of years, apart from disturbances produced by infrequent later impacts. Volcanic activity has been an important force in the shaping of the surfaces of the Moon and the terrestrial planets. Seismic activity on the Moon has been monitored by means of seismometers left on its surface by Apollo astronauts and by Lunokhod robotic rovers. Cratering on the largest scale seems to have ceased about three billion years ago, although on the Moon there is clear evidence for a continued cosmic drizzle of small particles, with the larger objects churning ("gardening") the lunar surface and the smallest producing microscopic impact pits in crystals in the lunar rocks.

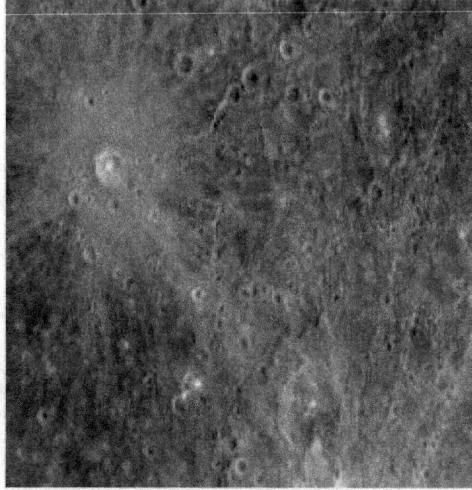

Mercury

Meteorite crater surrounded by rays of ejected material on Mercury, in a photograph taken by the Messenger probe, January 14, 2008. A chain of craters crosses the centre of the rayed crater.

All of the planets apart from the two closest to the Sun (Mercury and Venus) have natural satellites (moons) that are very diverse in appearance, size, and structure, as revealed in close-up observations from long-range space probes. The four outer dwarf planets have moons; Pluto has at least five moons, including one, Charon, fully half the size of Pluto itself. Over 200 asteroids and 80 Kuiper belt objects also have moons. Four planets (Jupiter, Saturn, Uranus, and Neptune), one dwarf planet (Haumea), and one Centaur object (Chariklo) have rings, disklike systems of small rocks and particles that orbit their parent bodies.

Charon

Pluto's moon Charon, in a photo taken by the New Horizons spacecraft, July 13, 2015.

Lunar Exploration

During the U.S. Apollo missions a total weight of 381.7 kg (841.5 pounds) of lunar material was collected; an additional 300 grams (0.66 pounds) was brought back by unmanned Soviet Luna vehicles. About 15 percent of the Apollo samples have been distributed for analysis, with the remainder stored at the NASA Johnson Space Center, Houston, Texas. The opportunity to employ a wide range of laboratory techniques on these lunar samples has revolutionized planetary science. The results of the analyses have enabled investigators to determine the composition and age of the lunar surface. Seismic observations have made it possible to probe the lunar interior. In addition, retroreflectors left on the Moon's surface by Apollo astronauts have allowed high-power laser beams to be sent from Earth to the Moon and back, permitting scientists to monitor the Earth-Moon distance to an accuracy of a few centimetres. This experiment, which has provided data used in calculations of the dynamics of the Earth-Moon system, has shown that the separation of the two bodies is increasing by 4.4 cm (1.7 inches) each year. (For additional information on lunar studies.)

Apollo 17 geologist-astronaut Harrison Schmitt at the foot of a huge split boulder, December 13, 1972, during the mission's third extravehicular exploration of the Taurus-Littrow Valley landing site.

Planetary Studies

Mercury is too hot to retain an atmosphere, but Venus's brilliant white appearance is the result of its being completely enveloped in thick clouds of carbon dioxide, impenetrable at visible wavelengths. Below the upper clouds, Venus has a hostile atmosphere containing clouds of sulfuric acid droplets.

The cloud cover shields the planet's surface from direct sunlight, but the energy that does filter through warms the surface, which then radiates at infrared wavelengths. The long-wavelength infrared radiation is trapped by the dense clouds such that an efficient greenhouse effect keeps the surface temperature near 465 °C (870 °F, 740 K). Radar, which can penetrate the thick Venusian clouds, has been used to map the planet's surface. In contrast, the atmosphere of Mars is very thin and is composed mostly of carbon dioxide (95 percent), with very little water vapour; the planet's surface pressure is only about 0.006 that of Earth. The outer planets have atmospheres composed largely of light gases, mainly hydrogen and helium.

Photo mosaic of Mercury, taken by
the Mariner 10 spacecraft, 1974.

Each planet rotates on its axis, and nearly all of them rotate in the same direction—counterclockwise as viewed from above the ecliptic. The two exceptions are Venus, which rotates in the clockwise direction beneath its cloud cover, and Uranus, which has its rotation axis very nearly in the plane of the ecliptic.

Some of the planets have magnetic fields. Earth's field extends outward until it is disturbed by the solar wind—an outward flow of protons and electrons from the Sun—which carries a magnetic field along with it. Through processes not yet fully understood, particles from the solar wind and galactic cosmic rays (high-speed particles from outside the solar system) populate two doughnut-shaped regions called the Van Allen radiation belts. The inner belt extends from about 1,000 to 5,000 km (600 to 3,000 miles) above Earth's surface, and the outer from roughly 15,000 to 25,000 km (9,300 to 15,500 miles). In these belts, trapped particles spiral along paths that take them around Earth while bouncing back and forth between the Northern and Southern hemispheres, with their orbits controlled by Earth's magnetic field. During periods of increased solar activity, these regions of trapped particles are disturbed, and some of the particles move down into Earth's atmosphere, where they collide with atoms and molecules to produce auroras.

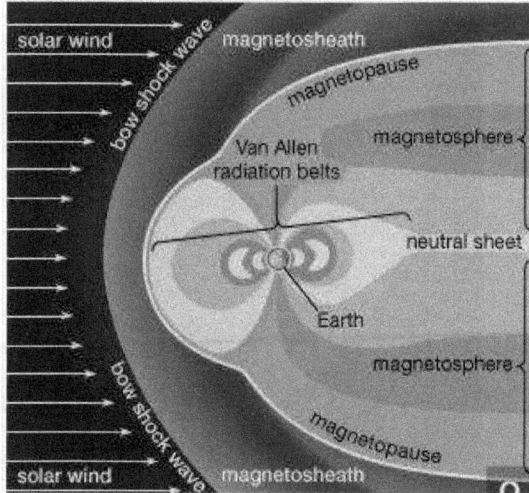

The Van Allen radiation belts contained within Earth's magnetosphere. Pressure from the solar wind is responsible for the asymmetrical shape of the magnetosphere and the belts.

Jupiter has a magnetic field far stronger than Earth's and many more trapped electrons, whose synchrotron radiation (electromagnetic radiation emitted by high-speed charged particles that are forced to move in curved paths, as under the influence of a magnetic field) is detectable from Earth. Bursts of increased radio emission are correlated with the position of Io, the innermost of the four Galilean moons of Jupiter. Saturn has a magnetic field that is much weaker than Jupiter's, but it too has a region of trapped particles. Mercury has a weak magnetic field that is only about 1 percent as strong as Earth's and shows no evidence of trapped particles. Uranus and Neptune have fields that are less than one-tenth the strength of Saturn's and appear much more complex than that of Earth. No field has been detected around Venus or Mars.

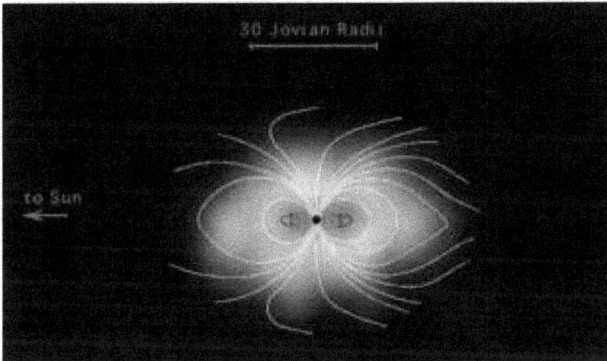

Jupiter's Magnetosphere

Jupiter's magnetosphere as observed by the Cassini spacecraft in 2000. The magnetosphere is the largest object in the solar system.

Investigations of the Smaller Bodies

More than 500,000 asteroids with well-established orbits are known, and thousands of additional objects are discovered each year. Hundreds of thousands more have been seen, but their orbits have not been as well determined. It is estimated that several million asteroids exist, but most are small, and their combined mass is estimated to be less than a thousandth that of Earth. Most of the asteroids have orbits close to the ecliptic and move in the asteroid belt, between 2.3 and 3.3 AU from the Sun. Because some asteroids travel in orbits that can bring them close to Earth, there is a possibility of a collision that could have devastating results (see Earth impact hazard).

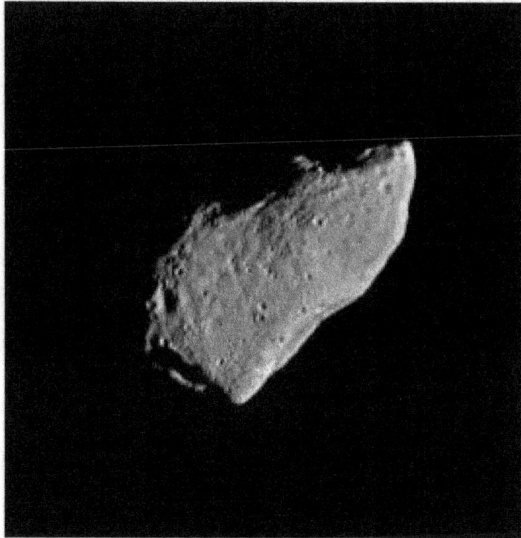

Gaspra

Asteroid Gaspra, composite of two images taken by the Galileo spacecraft on October 29, 1991. Galileo observed some 600 impact craters on Gaspra, from the large concavity visible on the lower right to craters as small as 100 metres (330 feet) in diameter. The asteroid's irregular shape and fracture lines suggest that it was once part of a larger body.

Comets are considered to come from a vast reservoir, the Oort cloud, orbiting the Sun at distances of 20,000–50,000 AU or more and containing trillions of icy objects—latent comet nuclei—with the potential to become active comets. Many comets have been observed over the centuries. Most make only a single pass through the inner solar system, but some are deflected by Jupiter or Saturn into orbits that allow them to return at predictable times. Halley's Comet is the best known of these periodic comets; its next return into the inner solar system is predicted for 2061. Many short-period comets are thought to come from the Kuiper belt, a region lying mainly between 30 AU and 50 AU from the Sun—beyond Neptune's orbit but including part of Pluto's—and housing perhaps hundreds of millions of comet nuclei. Very few comet masses have been well determined, but most are probably less than 10^{18} grams, one-billionth the mass of Earth.

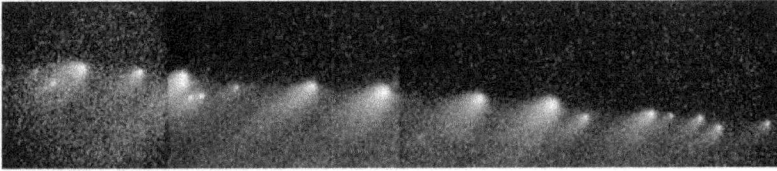

Composite image of the fragments of Comet Shoemaker-Levy 9, taken by the Hubble Space Telescope, 1994.

Since the 1990s more than a thousand comet nuclei in the Kuiper belt have been observed with large telescopes; a few are about half the size of Pluto, and Pluto is the largest Kuiper belt object. Pluto's orbital and physical characteristics had long caused it to be regarded as an anomaly among the planets. However, after the discovery of numerous other Pluto-like objects beyond Neptune, Pluto was seen to be no longer unique in its "neighbourhood" but rather a giant member of the local population. Consequently, in 2006 astronomers at the general assembly of the International Astronomical Union elected to create the new category of dwarf planets for objects with such qualifications. Pluto, Eris, and Ceres, the latter being the largest member of the asteroid belt, were given this distinction. Two other Kuiper belt objects, Makemake and Haumea, were also designated as dwarf planets.

Pluto

Pluto as seen by the New Horizons spacecraft, July 14, 2015.

Smaller than the observed asteroids and comets are the meteoroids, lumps of stony or metallic material believed to be mostly fragments of asteroids. Meteoroids vary from small rocks to boulders weighing a ton or more. A relative few have orbits that bring them into Earth's atmosphere and down to the surface as meteorites. Most meteorites that have been collected on Earth are probably from asteroids. A few have been identified as being from the Moon, Mars, or the asteroid Vesta.

Chelyabinsk meteorite of 2013
A cloud trail left behind by a meteorite that later exploded over Chelyabinsk province, Russia, February 15, 2013.

Meteorites are classified into three broad groups: stony (chondrites and achondrites; about 94 percent), iron (5 percent), and stony-iron (1 percent). Most meteoroids that enter the atmosphere heat up sufficiently to glow and appear as meteors, and the great majority of these vaporize completely or break up before they reach the surface. Many, perhaps most, meteors occur in showers (see meteor shower) and follow orbits that seem to be identical with those of certain comets, thus pointing to a cometary origin. For example, each May, when Earth crosses the orbit of Halley's Comet, the Eta Aquarid meteor shower occurs. Micrometeorites (interplanetary dust particles), the smallest meteoroidal particles, can be detected from Earth-orbiting satellites or collected by specially equipped aircraft flying in the stratosphere and returned for laboratory inspection. Since the late 1960s numerous meteorites have been found in the Antarctic on the surface of stranded ice flows (see Antarctic meteorites). Some meteorites contain microscopic crystals whose isotopic proportions are unique and appear to be dust grains that formed in the atmospheres of different stars.

Hoba meteorite, lying where it was discovered in 1920 in Grootfontein, Namibia. The object, the largest meteorite known and an iron meteorite by classification, is made of nickel-iron alloy and estimated to weigh nearly 60 tons.

Determinations of Age and Chemical Composition

The age of the solar system, taken to be close to 4.6 billion years, has been derived from measurements of radioactivity in meteorites, lunar samples, and Earth's crust. Abundances of isotopes of uranium, thorium, and rubidium and their decay products, lead and strontium, are the measured quantities.

Assessment of the chemical composition of the solar system is based on data from Earth, the Moon, and meteorites as well as on the spectral analysis of light from the Sun and planets. In broad outline, the solar system abundances of the chemical elements decrease with increasing atomic weight. Hydrogen atoms are by far the most abundant, constituting 91 percent; helium is next, with 8.9 percent; and all other types of atoms together amount to only 0.1 percent.

Theories of Origin

The origin of Earth, the Moon, and the solar system as a whole is a problem that has not yet been settled in detail. The Sun probably formed by condensation of the central region of a large cloud of gas and dust, with the planets and other bodies of the solar system forming soon after, their composition strongly influenced by the temperature and pressure gradients in the evolving solar nebula. Less-volatile materials could condense into solids relatively close to the Sun to form the terrestrial planets. The abundant, volatile lighter elements could condense only at much greater distances to form the giant gas planets.

Four recently formed stars in the Orion nebula show what the solar system might have looked like at its birth some 4.6 billion years ago. Gravity collapsed dense globules of dust and gas into each of the four stars. The outer material formed into the surrounding protoplanetary disks, which are believed to be the precursors of planets. The bright glow in each image is the star, and the dark oval is the protoplanetary disk seen at an angle.

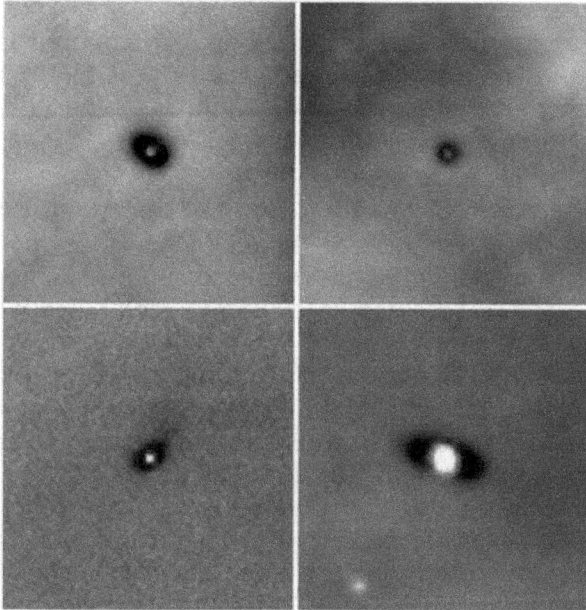

Four recently formed stars in the Orion nebula show what the solar system might have looked like at its birth some 4.6 billion years ago. Gravity collapsed dense globules of dust and gas into each of the four stars. The outer material formed into the surrounding protoplanetary disks, which are believed to be the precursors of planets. The bright glow in each image is the star, and the dark oval is the protoplanetary disk seen at an angle.

In the1990s astronomers confirmed that other stars have one or more planets revolving around them. Studies of these planetary systems have both supported and challenged astronomers' theoretical models of how Earth's solar system formed. Unlike the solar system, many extrasolar planetary systems have large gas giants like Jupiter orbiting very close to their stars, and in some cases these "hot Jupiters" are closer to their star than Mercury is to the Sun.

That so many gas giants, which form in the outer regions of their system, end up so close to their stars suggests that gas giants migrate and that such migration may have happened in the solar system's history. According to the Grand Tack hypothesis, Jupiter may have done so within a few million years of the solar system's formation. In this scenario, Jupiter is the first giant planet to form, at about 3 AU from the Sun. Drag from the protoplanetary disk causes it to fall inward to about 1.5 AU. However, by this time, Saturn begins to form at about 3 AU and captures Jupiter in a 3:2 resonance. (That is, for every three revolutions Jupiter makes, Saturn makes two.) The two planets migrate outward and clear away any material that would have gone to making Mars bigger. Mars should be bigger than Venus or Earth, but it is only half their size. The Grand Tack, in which Jupiter moves inward and then outward, explains Mars's small size.

About 500 million years after the Grand Tack, according to the Nice Model (named after the French city where it was first proposed), after the four giant planets—Jupiter, Saturn, Uranus, and Neptune—formed, they orbited 5–17 AU from the Sun. These planets were in a disk of smaller bodies called planetesimals and in orbital resonances with each other. About four billion years ago, gravitational interactions with the planetesimals increased the eccentricity of the planets' orbits, driving them out of resonance. Saturn, Uranus and Neptune migrated outward, and Jupiter migrated slightly inward. (Uranus and Neptune may even have switched places.) This migration scattered the disk, causing the Late Heavy Bombardment. The final remnant of the disk became the Kuiper belt.

The origin of the planetary satellites is not entirely settled. As to the origin of the Moon, the opinion of astronomers long oscillated between theories that saw its origin and condensation as simultaneous with the formation of Earth and those that posited a separate origin for the Moon and its later capture by Earth's gravitational field. Similarities and differences in abundances of the chemical elements and their isotopes on

Earth and the Moon challenged each group of theories. Finally, in the 1980s a model emerged that gained the support of most lunar scientists—that of a large impact on Earth and the expulsion of material that subsequently formed the Moon. For the outer planets, with their multiple satellites, many very small and quite unlike one another, the picture is less clear. Some of these moons have relatively smooth icy surfaces, whereas others are heavily cratered; at least one, Jupiter's Io, is volcanic. Some of the moons may have formed along with their parent planets, and others may have formed elsewhere and been captured.

Study of Extrasolar Planetary Systems

The first extrasolar planets were discovered in 1992, and more than 3,700 such planets are known. Over 600 of these systems have more than one planet. Because planets are much fainter than their stars, fewer than 100 have been imaged directly. Most extrasolar planets have been found through their transit, the small dimming of a star's light when a planet passes in front of it.

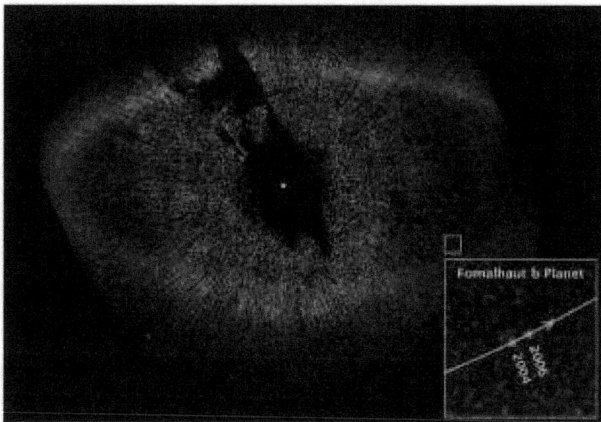

Fomalhaut b Planet

The extrasolar planet Fomalhaut b in images taken by the Hubble Space Telescope in 2004 and 2006. The black spot at the centre of the image is a coronagraph used to block the light from Fomalhaut, which is located at the white dot. The oval ring is Fomalhaut's dust belt, and the lines radiating from the centre of the image are scattered starlight.

Many of these planets are unlike those of the solar system. Hot Jupiters are large gas giants that orbit very close to their star. For example, HD 209458b is 0.69 times the mass of Jupiter and orbits its star every 3.52 days. Hot Neptunes are large ice giants about 10 percent of Jupiter's mass that also orbit very close to their star. Super-Earths are planets that are likely rocky like Earth but several times larger.

Artist's conception of the extrasolar planet HD 209458 b, some 150 light-years from Earth.

A primary goal of extrasolar planet research has been finding another planet that could support life. A useful guide for finding a life-supporting planet has been the concept of a habitable zone, the distance from a star where liquid water could survive on a planet's surface. About 40 planets have been found that are roughly Earth-sized and orbit in a habitable zone.

Kepler-452b

Study of the Stars
Measuring observable stellar properties

The measurable quantities in stellar astrophysics include the externally observable features of the stars: distance, temperature, radiation spectrum and luminosity, composition (of the outer layers), diameter, mass, and variability in any of these. Theoretical astrophysicists use these observations to model the structure of stars and to devise theories for their formation and evolution. Positional information can be used for dynamical analysis, which yields estimates of stellar masses.

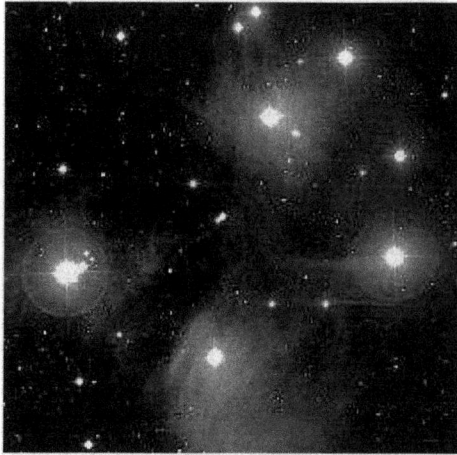

Pleiades
Bright nebulosity in the Pleiades (M45, NGC 1432),
distance 490 light-years. Cluster stars provide the light,
and surrounding clouds of dust reflect and
scatter the rays from the stars.

In a system dating back at least to the Greek astronomer-mathematician Hipparchus in the 2nd century BCE, apparent stellar brightness (m) is measured in magnitudes. Magnitudes are now defined such that a first-magnitude star is 100 times brighter than a star of sixth magnitude. The human eye cannot see stars fainter than about sixth magnitude, but modern instruments used with large telescopes can record stars as faint as about 30th magnitude. By convention, the absolute magnitude (M) is defined as the magnitude that a star would appear to have if it were located at a standard distance of 10 parsecs. These quantities are related through the expression m − M = 5 log10 r − 5, in which r is the star's distance in parsecs.

The magnitude scale is anchored on a group of standard stars. An absolute measure of radiant power is luminosity, which is related to the absolute magnitude and usually expressed in ergs per second (ergs/sec). (Sometimes the luminosity is stated in terms of the solar luminosity, 3.86×1033 ergs/sec.) Luminosity can be calculated when m and r are known. Correction might be necessary for the interstellar absorption of starlight.

There are several methods for measuring a star's diameter. From the brightness and distance, the luminosity (L) can be calculated, and, from observations of the brightness at different wavelengths, the temperature (T) can be calculated. Because the radiation from many stars can be well approximated by a Planck blackbody spectrum, these measured quantities can be related through the expression $L = 4\pi R^2 \sigma T^4$, thus providing a means of calculating R, the star's radius. In this expression, σ is the Stefan-Boltzmann constant, 5.67×10^{-5} ergs/cm^{2}K^{4}sec, in which K is the temperature in kelvins. (The radius R refers to the star's photosphere, the region where the star becomes effectively opaque to outside observation.) Stellar angular diameters can be measured through interferometry—that is, the combining of several telescopes together to form a larger instrument that can resolve sizes smaller than those that an individual telescope can resolve. Alternatively, the intensity of the starlight can be monitored during occultation by the Moon, which produces diffraction fringes whose pattern depends on the angular diameter of the star. Stellar angular diameters of several milliarcseconds can be measured.

Many stars occur in binary systems, in which the two partners orbit their mutual centre of mass. Such a system provides the best measurement of stellar masses. The period (P) of a binary system is related to the masses of the two stars (m1 and m2) and the orbital semimajor axis (mean radius; a) via Kepler's third law: $P2 = 4\pi 2a3/G(m1 + m2)$. (G is the universal gravitational constant.) From diameters and masses, average values of the stellar density can be calculated and thence the central pressure. With the assumption of an equation of state, the central temperature can then be calculated. For example, in the Sun the central density is 158 grams per cubic cm; the pressure is calculated to be more than one billion times the pressure of Earth's atmosphere at sea level and the temperature around 15 million K (27 million °F). At this temperature, all atoms are ionized, and so the solar interior consists of a plasma, an ionized gas with hydrogen nuclei (i.e., protons), helium nuclei, and electrons as major constituents. A small fraction of the hydrogen nuclei possess sufficiently high speeds that, on colliding, their electrostatic repulsion is overcome, resulting in the formation, by means of a set of fusion reactions, of helium nuclei and a release of energy. Some of this energy is carried away by neutrinos, but most of it is carried by photons to the surface of the Sun to maintain its luminosity.

Other stars, both more and less massive than the Sun, have broadly similar structures, but the size, central pressure and temperature, and fusion rate are functions of the star's mass and composition. The stars and their internal fusion (and resulting luminosity) are held stable against collapse through a delicate balance between the inward pressure produced by gravitational attraction and the outward pressure supplied by the photons produced in the fusion reactions.

Stars that are in this condition of hydrostatic equilibrium are termed main-sequence stars, and they occupy a well-defined band on the Hertzsprung-Russell (H-R) diagram, in which luminosity is plotted against colour index or temperature. Spectral classification, based initially on the colour index, includes the major spectral types O, B, A, F, G, K and M, each subdivided into 10 parts. Temperature is deduced from broadband spectral measurements in several standard wavelength intervals. Measurement of apparent magnitudes in two spectral regions, the B and V bands (centred on 4350 and 5550 angstroms, respectively), permits calculation of the colour index, $CI = mB - mV$, from which the temperature can be calculated.

Hertzsprung-Russell diagram

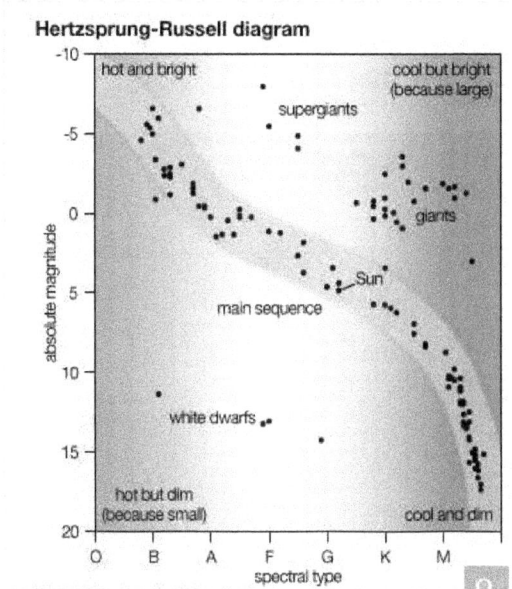

Hertzsprung-Russell diagram
Hertzsprung-Russell diagram. Spectral type
(a measure of a star's temperature), following the order
introduced by American astronomer Annie Jump Cannon, is
plotted on the horizontal axis, and absolute magnitude (the
intrinsic brightness of a star) is plotted on the vertical axis.

For a given temperature, there are stars that are much
more luminous than main-sequence stars. Given the
dependence of luminosity on the square of the radius and the
fourth power of the temperature (R^2T^4 of the luminosity
expression above), greater luminosity implies larger radius, and
such stars are termed giant stars or supergiant stars.
Conversely, stars with luminosities much less than those of
main-sequence stars of the same temperature must be smaller
and are termed white dwarf stars. Surface temperatures of
white dwarfs typically range from 10,000 to 12,000 K (18,000
to 21,000 °F), and they appear visually as white or blue-white.

The strength of spectral lines of the more abundant
elements in a star's atmosphere allows additional subdivisions
within a class. Thus, the Sun, a main-sequence star, is
classified as G2 V, in which the V denotes main sequence.
Betelgeuse, a red giant with a surface temperature about half
that of the Sun but with a luminosity of about 10,000 solar

units, is classified as M2 Iab. In this classification, the spectral type is M2, and the Iab indicates a giant, well above the main sequence on the H-R diagram.

Betelgeuse
Betelgeuse imaged in ultraviolet light
by the Hubble Space Telescope.

Star Formation and Evolution

The range of physically allowable masses for stars is very narrow. If the star's mass is too small, the central temperature will be too low to sustain fusion reactions. The theoretical minimum stellar mass is about 0.08 solar mass. An upper theoretical bound called the Eddington limit, of several hundred solar masses, has been suggested, but this value is not firmly defined. Stars as massive as this will have luminosities about one million times greater than that of the Sun.

Horsehead Nebula

A general model of star formation and evolution has been developed, and the major features seem to be established. A large cloud of gas and dust can contract under its own gravitational attraction if its temperature is sufficiently low. As gravitational energy is released, the contracting central material heats up until a point is reached at which the outward radiation pressure balances the inward gravitational pressure, and contraction ceases. Fusion reactions take over as the star's primary source of energy, and the star is then on the main sequence. The time to pass through these formative stages and onto the main sequence is less than 100 million years for a star with as much mass as the Sun. It takes longer for less massive stars and a much shorter time for those much more massive.

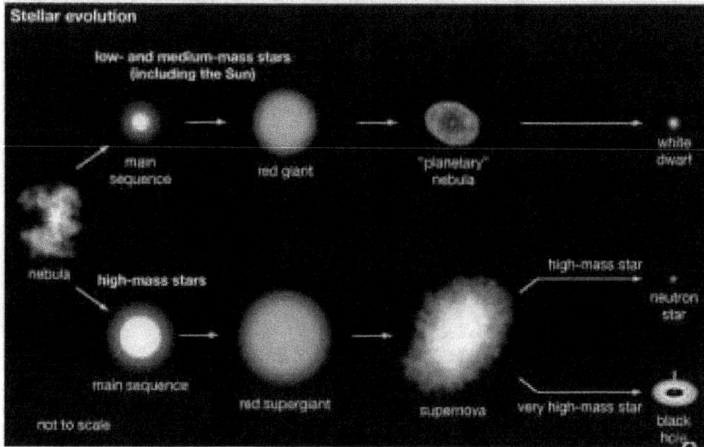

Stellar Evolution

Once a star has reached its main-sequence stage, it evolves relatively slowly, fusing hydrogen nuclei in its core to form helium nuclei. Continued fusion not only releases the energy that is radiated but also results in nucleosynthesis, the production of heavier nuclei.

Stellar evolution has of necessity been followed through computer modeling, because the timescales for most stages are generally too extended for measurable changes to be observed, even over a period of many years. One exception is the supernova, the violently explosive finale of certain stars. Different types of supernovas can be distinguished by their spectral lines and by changes in luminosity during and after the outburst. In Type Ia, a white dwarf star attracts matter from a nearby companion; when the white dwarf's mass exceeds about 1.4 solar masses, the star implodes and is completely destroyed. Type II supernovas are not as luminous as Type Ia and are the final evolutionary stage of stars more massive than about eight solar masses. Type Ib and Ic supernovas are like Type II in that they are from the collapse of a massive star, but they do not retain their hydrogen envelope.

The nature of the final products of stellar evolution depends on stellar mass. Some stars pass through an unstable stage in which their dimensions, temperature, and luminosity change cyclically over periods of hours or days. These so-called Cepheid variables serve as standard candles for distance measurements. Some stars blow off their outer layers to produce planetary nebulas. The expanding material can be seen glowing in a thin shell as it disperses into the interstellar medium while the remnant core, initially with a surface temperature as high as 100,000 K (180,000 °F), cools to become a white dwarf. The maximum stellar mass that can exist as a white dwarf is about 1.4 solar masses and is known as the Chandrasekhar limit. More-massive stars may end up as either neutron stars or black holes.

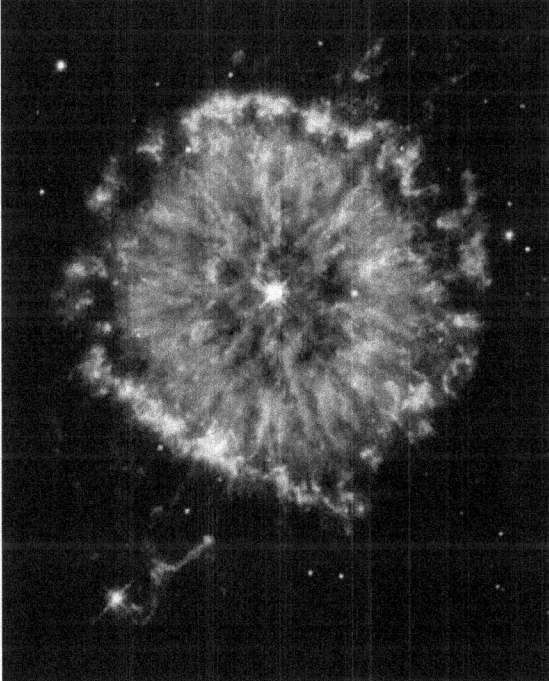

Nebula NGC 6751 formed several thousands of years ago as a dying star in the constellation Aquila threw off its outer layers of gas, which then fluoresced into a shell. The image, captured via NASA's Hubble Space Telescope, combines views taken through three different color filters in order to show the different temperatures of the nebular gases, with blue representing the hottest and orange and red the coolest. Our own sun will undergo a similar process in about 6 billion years, as it nears the end of its life.

The average density of a white dwarf is calculated to exceed one million grams per cubic cm. Further compression is limited by a quantum condition called degeneracy, in which only certain energies are allowed for the electrons in the star's interior. Under sufficiently great pressure, the electrons are forced to combine with protons to form neutrons. The resulting neutron star will have a density in the range of 10^{14}–10^{15} grams per cubic cm, comparable to the density within atomic nuclei. The behaviour of large masses having nuclear densities is not yet sufficiently understood to be able to set a limit on the maximum size of a neutron star, but it is thought to be less than three solar masses.

Still more-massive remnants of stellar evolution would have smaller dimensions and would be even denser that neutron stars. Such remnants are conceived to be black holes, objects so compact that no radiation can escape from within a characteristic distance called the Schwarzschild radius. This critical dimension is defined by Rs = 2GM/c2. (Rs is the Schwarzschild radius, G is the gravitational constant, M is the object's mass, and c is the speed of light.) For an object of three solar masses, the Schwarzschild radius would be about three kilometres. Radiation emitted from beyond the Schwarzschild radius can still escape and be detected.

Although no light can be detected coming from within a black hole, the presence of a black hole may be manifested through the effects of its gravitational field, as, for example, in a binary star system. If a black hole is paired with a normal visible star, it may pull matter from its companion toward itself. This matter is accelerated as it approaches the black hole and becomes so intensely heated that it radiates large amounts of X-rays from the periphery of the black hole before reaching the Schwarzschild radius. Some candidates for stellar black holes have been found—e.g., the X-ray source Cygnus X-1. Each of them has an estimated mass clearly exceeding that allowable for a neutron star, a factor crucial in the identification of possible black holes. (Supermassive black holes that do not originate as individual stars are thought to exist at the centre of active galaxies; see below Study of other galaxies and related phenomena.)

Cygnus X-1
the X-ray source Cygnus X-1.
A black hole pulls in matter from a stellar companion.

Whereas the existence of stellar black holes has been strongly indicated, the existence of neutron stars was confirmed in 1968 when they were identified with the then newly discovered pulsars, objects characterized by the emission of radiation at short and extremely regular intervals, generally between 1 and 1,000 pulses per second and stable to better than a part per billion. Pulsars are considered to be rotating neutron stars, remnants of some supernovas.

Study of the Milky Way Galaxy

Stars are not distributed randomly throughout space. Many stars are in systems consisting of two or three members separated by less than 1,000 AU. On a larger scale, star clusters may contain many thousands of stars. Galaxies are much larger systems of stars and usually include clouds of gas and dust.

Milky Way Galaxy as seen from Earth

The solar system is located within the Milky Way Galaxy, close to its equatorial plane and about 8 kiloparsecs from the galactic centre. The galactic diameter is about 30 kiloparsecs, as indicated by luminous matter. There is evidence, however, for nonluminous matter—so-called dark matter—extending out nearly twice this distance. The entire system is rotating such that, at the position of the Sun, the orbital speed is about 220 km per second (almost 500,000 miles per hour) and a complete circuit takes roughly 240 million years. Application of Kepler's third law leads to an estimate for the galactic mass of about 100 billion solar masses. The rotational velocity can be measured from the Doppler shifts observed in the 21-cm emission line of neutral hydrogen and the lines of millimetre wavelengths from various molecules, especially carbon monoxide. At great distances from the galactic centre, the rotational velocity does not drop off as expected but rather increases slightly. This behaviour appears to require a much larger galactic mass than can be accounted for by the known (luminous) matter. Additional evidence for the presence of dark matter comes from a variety of other observations. The nature and extent of the dark matter (or missing mass) constitutes one of today's major astronomical puzzles.

There are about 100 billion stars in the Milky Way Galaxy. Star concentrations within the galaxy fall into three types: open clusters, globular clusters, and associations. Open clusters lie primarily in the disk of the galaxy; most contain between 50 and 1,000 stars within a region no more than 10 parsecs in diameter. Stellar associations tend to have somewhat fewer stars; moreover, the constituent stars are not as closely grouped as those in the clusters and are for the most part hotter. Globular clusters, which are widely scattered around the galaxy, may extend up to about 100 parsecs in diameter and may have as many as a million stars. The importance to astronomers of globular clusters lies in their use as indicators of the age of the galaxy. Because massive stars evolve more rapidly than do smaller stars, the age of a cluster can be estimated from its H-R diagram. In a young cluster the main sequence will be well populated, but in an old cluster the heavier stars will have evolved away from the main sequence. The extent of the depopulation of the main sequence provides an index of age. In this way, the oldest globular clusters have been found to be about 12.5 billion years old, which should therefore be the minimum age for the galaxy.

Globular cluster M80 (also known as NGC 6093) in an optical image taken by the Hubble Space Telescope. M80 is located 28,000 light-years from Earth and contains hundreds of thousands of stars.

Investigations of Interstellar Matter

The interstellar medium, composed primarily of gas and dust, occupies the regions between the stars. On average, it contains less than one atom in each cubic centimetre, with about 1 percent of its mass in the form of minute dust grains. The gas, mostly hydrogen, has been mapped by means of its 21-cm emission line. The gas also contains numerous molecules. Some of these have been detected by the visible-wavelength absorption lines that they impose on the spectra of more-distant stars, while others have been identified by their own emission lines at millimetre wavelengths. Many of the interstellar molecules are found in giant molecular clouds, wherein complex organic molecules have been discovered.

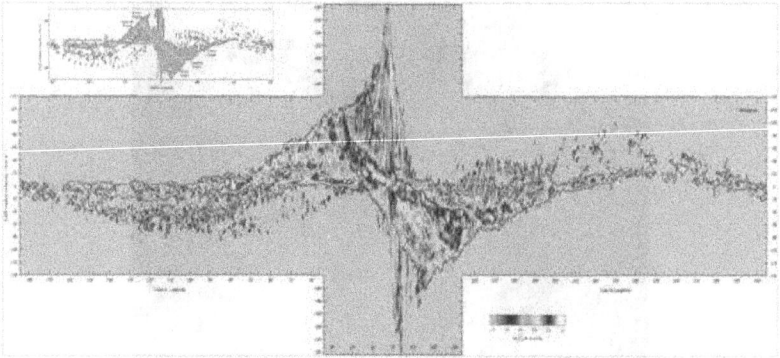

Longitude-velocity map of the Milky Way Galaxy as shown by spectral line emission of carbon monoxide in molecular clouds. The vertical axis represents velocity and the horizontal axis longitude. The gentle curves in the left and right portions of the map trace the spiral arms of the Milky Way Galaxy. The vertical structure in the middle of the map is the centre of the Galaxy. The emission stretching from the upper left to the lower right in the middle portion of the map is the "molecular ring," a ring of gas and dust in orbit between 4 and 8 kiloparsecs from the centre of the Galaxy.

In the vicinity of a very hot O- or B-type star, the intensity of ultraviolet radiation is sufficiently high to ionize the surrounding hydrogen out to a distance as great as 100 parsecs to produce an H II region, known as a Strömgren sphere. Such regions are strong and characteristic emitters of radiation at radio wavelengths, and their dimensions are well calibrated in terms of the luminosity of the central star. Using radio interferometers, astronomers are able to measure the angular diameters of H II regions even in some external galaxies and can thereby deduce the great distances to those remote systems. This method can be used for distances up to about 30 megaparsecs. (For additional information on H II regions, see nebula: Diffuse nebulae (H II regions).)

H II region
NGC 604, H II region inside the Triangulum
Galaxy (M33), photographed by Hubble Space Telescope. NGC
604 is one of the largest H II regions; it contains more than
200 stars and is about 1,500 light-years in diameter.

Interstellar dust grains scatter and absorb starlight, the effect being roughly inversely proportional to wavelength from the infrared to the near ultraviolet. As a result, stellar spectra tend to be reddened. Absorption typically amounts to about one magnitude per kiloparsec but varies considerably in different directions. Some dusty regions contain silicate materials, identified by a broad absorption feature around a wavelength of 10 μm. Other prominent spectral features in the infrared range have been sometimes, but not conclusively, attributed to graphite grains and polycyclic aromatic hydrocarbons (PAHs).

Starlight often shows a small degree of polarization (a few percent), with the effect increasing with stellar distance. This is attributed to the scattering of the starlight from dust grains that have been partially aligned in a weak interstellar magnetic field. The strength of this field is estimated to be a few microgauss, very close to the strength inferred from observations of nonthermal cosmic radio noise. This radio background has been identified as synchrotron radiation, emitted by cosmic-ray electrons traveling at nearly the speed of light and moving along curved paths in the interstellar magnetic field. The spectrum of the cosmic radio noise is close to what is calculated on the basis of measurements of the cosmic rays near Earth.

Cosmic rays constitute another component of the interstellar medium. Cosmic rays that are detected in the vicinity of Earth comprise high-speed nuclei and electrons. Individual particle energies, expressed in electron volts (eV; 1 eV = 1.6×10^{-12} erg), range with decreasing numbers from about 10^6 eV to more than 10^{20} eV. Among the nuclei, hydrogen nuclei are the most plentiful at 86 percent, helium nuclei next at 13 percent, and all other nuclei together at about 1 percent. Electrons are about 2 percent as abundant as the nuclear component. (The relative numbers of different nuclei vary somewhat with kinetic energy, while the electron proportion is strongly energy-dependent.)

A minority of cosmic rays detected in Earth's vicinity are produced in the Sun, especially at times of increased solar activity (as indicated by sunspots and solar flares). The origin of galactic cosmic rays has not yet been conclusively identified, but they are thought to be produced in stellar processes such as supernova explosions, perhaps with additional acceleration occurring in the interstellar regions. (For additional information on interstellar matter, see Milky Way Galaxy: The general interstellar medium.)

Observations of the Galactic Centre
The central region of the Milky Way Galaxy is so heavily obscured by dust that direct observation has become possible only with the development of astronomy at nonvisual wavelengths—namely, radio, infrared, and, more recently, X-ray and gamma-ray wavelengths. Together, these observations have revealed a nuclear region of intense activity, with a large number of separate sources of emission and a great deal of dust. Detection of gamma-ray emission at a line energy of 511,000 eV, which corresponds to the annihilation of electrons and positrons (the antimatter counterpart of electrons), along with radio mapping of a region no more than 20 AU across, points to a very compact and energetic source, designated Sagittarius A*, at the centre of the galaxy. Sagittarius A* is a supermassive black hole with a mass equivalent to 4,310,000 Suns.

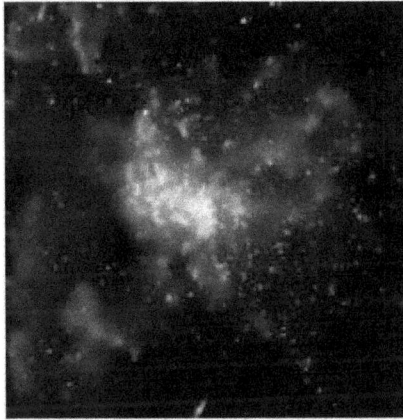

Cosmic radio-wave source Sagittarius A* in an image from the Chandra X-ray Observatory. Sagittarius A* is an extremely bright source within the larger Sagittarius A complex and contains the black hole at the Milky Way Galaxy's centre.

Study of Other Galaxies and Related Phenomena

Galaxies are normally classified into three principal types according to their appearance: spiral, elliptical, and irregular. Galactic diameters are typically in the tens of kiloparsecs and the distances between galaxies typically in megaparsecs.

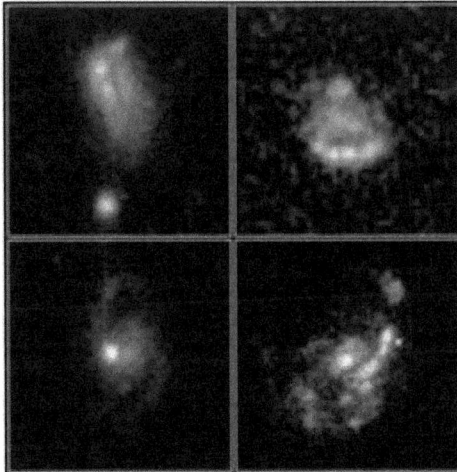

Four irregular galaxies, as observed by the Hubble Space Telescope.

Spiral galaxies—of which the Milky Way system is a characteristic example—tend to be flattened, roughly circular systems with their constituent stars strongly concentrated along spiral arms. These arms are thought to be produced by traveling density waves, which compress and expand the galactic material. Between the spiral arms exists a diffuse interstellar medium of gas and dust, mostly at very low temperatures (below 100 K [-280 °F, -170 °C]). Spiral galaxies are typically a few kiloparsecs in thickness; they have a central bulge and taper gradually toward the outer edges.

Whirlpool Galaxy (M51); NGC 5195
The Whirlpool Galaxy (left), also known as M51,
an Sc galaxy accompanied by a small irregular companion
galaxy, NGC 5195 (right).

Ellipticals show none of the spiral features but are more densely packed stellar systems. They range in shape from nearly spherical to very flattened and contain little interstellar matter. Irregular galaxies number only a few percent of all stellar systems and exhibit none of the regular features associated with spirals or ellipticals.

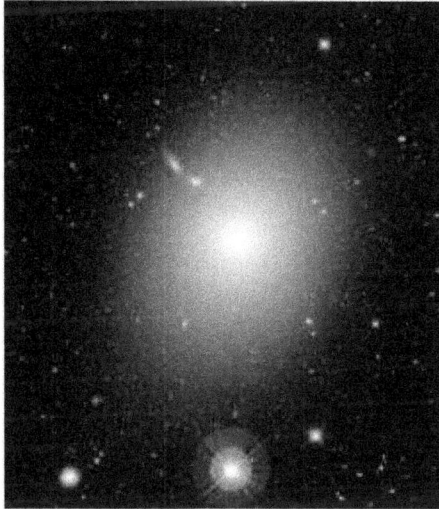

The giant elliptical galaxy M87, also known as Virgo A, in an optical image taken by the Canada-France-Hawaii Telescope on Mauna Kea, Hawaii. M87 appears near the centre of the Virgo Cluster of galaxies.

Properties vary considerably among the different types of galaxies. Spirals typically have masses in the range of a billion to a trillion solar masses, with ellipticals having values from 10 times smaller to 10 times larger and the irregulars generally 10–100 times smaller. Visual galactic luminosities show similar spreads among the three types, but the irregulars tend to be less luminous. In contrast, at radio wavelengths the maximum luminosity for spirals is usually 100,000 times less than for ellipticals or irregulars.

Quasars are objects whose spectra display very large redshifts, thus implying (in accordance with the Hubble law) that they lie at the greatest distances. They were discovered in 1963 but remained enigmatic for many years. They appear as starlike (i.e., very compact) sources of radio waves—hence their initial designation as quasi-stellar radio sources, a term later shortened to quasars. They are now considered to be the exceedingly luminous cores of distant galaxies. These energetic cores, which emit copious quantities of X-rays and gamma rays, are termed active galactic nuclei (AGN) and include the object Cygnus A and the nuclei of a class of galaxies called Seyfert galaxies. They may be powered by the infall of matter into supermassive black holes.

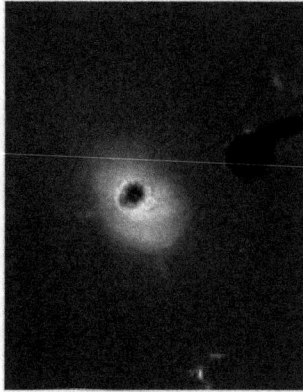

Quasar
3C 273, the brightest quasar, photographed by
the Hubble Space Telescope's Advanced Camera for Surveys.
The black region at the centre of the image is blocking light
from the central object, revealing the host galaxy of 3C 273.

The Milky Way Galaxy is one of the Local Group of galaxies, which contains about four dozen members and extends over a volume about two megaparsecs in diameter. Two of the closest members are the Magellanic Clouds, irregular galaxies about 50 kiloparsecs away. At about 740 kiloparsecs, the Andromeda Galaxy is one of the most distant in the Local Group. Some members of the group are moving toward the Milky Way system while others are traveling away from it. At greater distances, all galaxies are moving away from the Milky Way Galaxy. Their speeds (as determined from the redshifted wavelengths in their spectra) are generally proportional to their distances. The Hubble law relates these two quantities. In the absence of any other method, the Hubble law continues to be used for distance determinations to the farthest objects—that is, galaxies and quasars for which redshifts can be measured.

The Andromeda Galaxy, also known as the
Andromeda Nebula or M31. It is the closest spiral galaxy to
Earth, at a distance of 2.48 million light-years.

Cosmology

Cosmology is the scientific study of the universe as a unified
whole, from its earliest moments through its evolution to its
ultimate fate. The currently accepted cosmological model is the
big bang. In this picture, the expansion of the universe started
in an intense explosion 13.8 billion years ago. In this primordial
fireball, the temperature exceeded one trillion K, and most of
the energy was in the form of radiation. As the expansion
proceeded (accompanied by cooling), the role of the radiation
diminished, and other physical processes dominated in turn.
Thus, after about three minutes, the temperature had dropped
to the one-billion-K range, making it possible for nuclear
reactions of protons to take place and produce nuclei of
deuterium and helium. (At the higher temperatures that
prevailed earlier, these nuclei would have been promptly
disrupted by high-energy photons.) With further expansion, the
time between nuclear collisions had increased and the
proportion of deuterium and helium nuclei had stabilized. After
a few hundred thousand years, the temperature must have
dropped sufficiently for electrons to remain attached to nuclei
to constitute atoms. Galaxies are thought to have begun
forming after a few million years, but this stage is very poorly
understood. Star formation probably started much later, after
at least a billion years, and the process continues today.

Observational support for this general model comes from
several independent directions. The expansion has been

documented by the redshifts observed in the spectra of galaxies. Furthermore, the radiation left over from the original fireball would have cooled with the expansion. Confirmation of this relic energy came in 1965 with one of the most striking cosmic discoveries of the 20th century—the observation, at short radio wavelengths, of a widespread cosmic radiation corresponding to a temperature of almost 3 K (about −270 °C [−454 °F]). The shape of the observed spectrum is an excellent fit with the theoretical Planck blackbody spectrum. (The present best value for this temperature is 2.735 K, but it is still called three-degree radiation or the cosmic microwave background.) The spectrum of this cosmic radio noise peaks at approximately a one-millimetre wavelength, which is in the far infrared, a difficult region to observe from Earth; however, the spectrum has been well mapped by the Cosmic Background Explorer (COBE), Wilkinson Microwave Anisotropy Probe, and Planck satellites. Additional support for the big bang theory comes from the observed cosmic abundances of deuterium and helium. Normal stellar nucleosynthesis cannot produce their measured quantities, which fit well with calculations of production during the early stages of the big bang.

Early surveys of the cosmic background radiation indicated that it is extremely uniform in all directions (isotropic). Calculations have shown that it is difficult to achieve this degree of isotropy unless there was a very early and rapid inflationary period before the expansion settled into its present mode. Nevertheless, the isotropy posed problems for models of galaxy formation. Galaxies originate from turbulent conditions that produce local fluctuations of density, toward which more matter would then be gravitationally attracted. Such density variations were difficult to reconcile with the isotropy required by observations of the 3 K radiation. This problem was solved when the COBE satellite was able to detect the minute fluctuations in the cosmic background from which the galaxies formed.

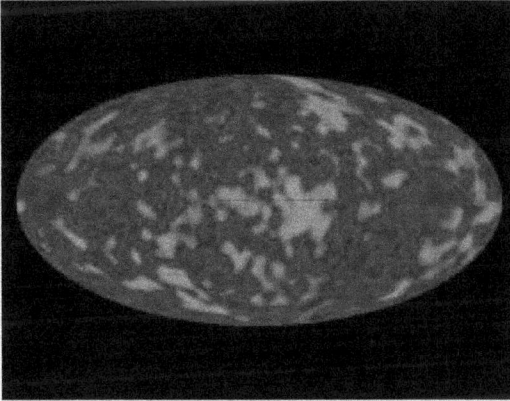

Image of the cosmic microwave background, taken by the Differential Microwave Radiometer on board the U.S. satellite Cosmic Background Explorer. The red features in the image show places where the universe was slightly denser, thus stimulating gravitational separation and, ultimately, the formation of galaxies.

The very earliest stages of the big bang are less well understood. The conditions of temperature and pressure that prevailed prior to the first microsecond require the introduction of theoretical ideas of subatomic particle physics. Subatomic particles are usually studied in laboratories with giant accelerators, but the region of particle energies of potential significance to the question at hand lies beyond the range of accelerators currently available. Fortunately, some important conclusions can be drawn from the observed cosmic helium abundance, which is dependent on conditions in the early big bang. The observed helium abundance sets a limit on the number of families of certain types of subatomic particles that can exist.

The age of the universe can be calculated in several ways. Assuming the validity of the big bang model, one attempts to answer the question: How long has the universe been expanding in order to have reached its present size? The numbers relevant to calculating an answer are Hubble's constant (i.e., the current expansion rate), the density of matter in the universe, and the cosmological constant, which allows for change in the expansion rate. In 2003 a calculation based on a fresh determination of Hubble's constant yielded an age of 13.7 billion ± 200 million years, although the precise value depends on certain assumed details of the model used. Independent estimates of stellar ages have yielded values less than this, as would be expected, but other estimates, based on supernova distance measurements, have arrived at values of about 15 billion years, still consistent, within the errors. In the big bang model the age is proportional to the reciprocal of Hubble's constant, hence the importance of determining H as reliably as possible. For example, a value for H of 100 km/sec/Mpc would lead to an age less than that of many stars, a physically unacceptable result.

A small minority of astronomers have developed alternative cosmological theories that are seriously pursued. The overwhelming professional opinion, however, continues to support the big bang model.

Finally, there is the question of the future behaviour of the universe: Is it open? That is to say, will the expansion continue indefinitely? Or is it closed, such that the expansion will slow down and eventually reverse, resulting in contraction? (The final collapse of such a contracting universe is sometimes termed the "big crunch.") The density of the universe seems to be at the critical density; that is, the universe is neither open nor closed but "flat." So-called dark energy, a kind of repulsive force that is now believed to be a major component of the universe, appears to be the decisive factor in predictions of the long-term fate of the cosmos. If this energy is a cosmological constant (as proposed in 1917 by Albert Einstein to correct certain problems in his model of the universe), then the result would be a "big chill." In this scenario, the universe would continue to expand, but its density would decrease. While old stars would burn out, new stars would no longer form. The universe would become cold and dark. The dark (nonluminous) matter component of the universe, whose composition remains unknown, is not considered sufficient to close the universe and cause it to collapse; it now appears to contribute only a fourth of the density needed for closure.

An additional factor in deciding the fate of the universe might be the mass of neutrinos. For decades the neutrino had been postulated to have zero mass, although there was no compelling theoretical reason for this to be so. From the observation of neutrinos generated in the Sun and other celestial sources such as supernovas, in cosmic-ray interactions with Earth's atmosphere, and in particle accelerators, investigators have concluded that neutrinos have some mass, though only an extremely small fraction of the mass of an electron. Although there are vast numbers of neutrinos in the universe, the sum of such small neutrino masses appears insufficient to close the universe.

*****

Relativity

Both Quantum mechanics and general relativity are theories of information. To know how the basics of information underlies both of them, we to think beyond shanon's classical information theory and thermodynamics and explore the general relativity and quantum theory to see how information lies beneath the universe as the driving factor.

Relativity, wide-ranging physical theories formed by the German-born physicist Albert Einstein. With his theories of special relativity (1905) and general relativity (1915), Einstein overthrew many assumptions underlying earlier physical theories, redefining in the process the fundamental concepts of space, time, matter, energy, and gravity. Along with quantum mechanics, relativity is central to modern physics. In particular, relativity provides the basis for understanding cosmic processes and the geometry of the universe itself.

"Special relativity" is limited to objects that are moving with respect to inertial frames of reference—i.e, in a state of uniform motion with respect to one another such that an observer cannot, by purely mechanical experiments, distinguish one from the other. Beginning with the behaviour of light (and all other electromagnetic radiation), the theory of special relativity draws conclusions that are contrary to everyday experience but fully confirmed by experiments. Special relativity revealed that the speed of light is a limit that can be approached but not reached by any material object; it is the origin of the most famous equation in science, $E = mc^2$; and it has led to other tantalizing outcomes, such as the "twin paradox."

"General relativity" is concerned with gravity, one of the fundamental forces in the universe. (The others are electricity and magnetism, which have been unified as electromagnetism, the strong force, and the weak force.) Gravity defines macroscopic behaviour, and so general relativity describes large-scale physical phenomena such as planetary dynamics, the birth and death of stars, black holes, and the evolution of the universe.

Special and general relativity have profoundly affected physical science and human existence, most dramatically in applications of nuclear energy and nuclear weapons. Additionally, relativity and its rethinking of the fundamental categories of space and time have provided a basis for certain philosophical, social, and artistic interpretations that have influenced human culture in different ways.

Cosmology Before Relativity
The Mechanical Universe

Relativity changed the scientific conception of the universe, which began in efforts to grasp the dynamic behaviour of matter. In Renaissance times, the great Italian physicist Galileo Galilei moved beyond Aristotle's philosophy to introduce the modern study of mechanics, which requires quantitative measurements of bodies moving in space and time. His work and that of others led to basic concepts, such as velocity, which is the distance a body covers in a given direction per unit time; acceleration, the rate of change of velocity; mass, the amount of material in a body; and force, a push or pull on a body.

The next major stride occurred in the late 17th century, when the British scientific genius Isaac Newton formulated his three famous laws of motion, the first and second of which are of special concern in relativity. Newton's first law, known as the law of inertia, states that a body that is not acted upon by external forces undergoes no acceleration—either remaining at rest or continuing to move in a straight line at constant speed. Newton's second law states that a force applied to a body changes its velocity by producing an acceleration that is proportional to the force and inversely proportional to the mass of the body. In constructing his system, Newton also defined space and time, taking both to be absolutes that are unaffected by anything external. Time, he wrote, "flows equably," while space "remains always similar and immovable."

Newton's laws proved valid in every application, as in calculating the behaviour of falling bodies, but they also provided the framework for his landmark law of gravity (the term, derived from the Latin gravis, or "heavy," had been in use since at least the 16th century). Beginning with the (perhaps mythical) observation of a falling apple and then considering the Moon as it orbits Earth, Newton concluded that an invisible force acts between the Sun and its planets. He formulated a comparatively simple mathematical expression for the gravitational force; it states that every object in the universe attracts every other object with a force that operates through empty space and that varies with the masses of the objects and the distance between them.

The law of gravity was brilliantly successful in explaining the mechanism behind Kepler's laws of planetary motion, which the German astronomer Johannes Kepler had formulated at the beginning of the 17th century. Newton's mechanics and law of gravity, along with his assumptions about the nature of space and time, seemed wholly successful in explaining the dynamics of the universe, from motion on Earth to cosmic events.

Light and the Ether

However, this success at explaining natural phenomena came to be tested from an unexpected direction—the behaviour of light, whose intangible nature had puzzled philosophers and scientists for centuries. In 1865 the Scottish physicist James Clerk Maxwell showed that light is an electromagnetic wave with oscillating electrical and magnetic components. Maxwell's equations predicted that electromagnetic waves would travel through empty space at a speed of almost exactly 3×10^8 metres per second (186,000 miles per second)—i.e., according with the measured speed of light. Experiments soon confirmed the electromagnetic nature of light and established its speed as a fundamental parameter of the universe.

Maxwell's remarkable result answered long-standing questions about light, but it raised another fundamental issue: if light is a moving wave, what medium supports it? Ocean waves and sound waves consist of the progressive oscillatory motion of molecules of water and of atmospheric gases, respectively. But what is it that vibrates to make a moving light wave? Or to put it another way, how does the energy embodied in light travel from point to point?

For Maxwell and other scientists of the time, the answer was that light traveled in a hypothetical medium called the ether (aether). Supposedly, this medium permeated all space without impeding the motion of planets and stars; yet it had to be more rigid than steel so that light waves could move through it at high speed, in the same way that a taut guitar string supports fast mechanical vibrations. Despite this contradiction, the idea of the ether seemed essential—until a definitive experiment disproved it.

In 1887 the German-born American physicist A.A. Michelson and the American chemist Edward Morley made exquisitely precise measurements to determine how Earth's motion through the ether affected the measured speed of light. In classical mechanics, Earth's movement would add to or subtract from the measured speed of light waves, just as the speed of a ship would add to or subtract from the speed of ocean waves as measured from the ship. But the Michelson-Morley experiment had an unexpected outcome, for the measured speed of light remained the same regardless of Earth's motion. This could only mean that the ether had no meaning and that the behaviour of light could not be explained by classical physics. The explanation emerged, instead, from Einstein's theory of special relativity.

Special Relativity
Einstein's Gedanken Experiments
Scientists such as Austrian physicist Ernst Mach and French mathematician Henri Poincaré had critiqued classical mechanics or contemplated the behaviour of light and the meaning of the ether before Einstein. Their efforts provided a background for Einstein's unique approach to understanding the universe, which he called in his native German a Gedankenexperiment, or "thought experiment."

Einstein described how at age 16 he watched himself in his mind's eye as he rode on a light wave and gazed at another light wave moving parallel to his. According to classical physics, Einstein should have seen the second light wave moving at a relative speed of zero. However, Einstein knew that Maxwell's electromagnetic equations absolutely require that light always move at 3×10^8 metres per second in a vacuum. Nothing in the theory allows a light wave to have a speed of zero. Another problem arose as well: if a fixed observer sees light as having a speed of 3×10^8 metres per second, whereas an observer moving at the speed of light sees light as having a speed of zero, it would mean that the laws of electromagnetism depend on the observer. But in classical mechanics the same laws apply for all observers, and Einstein saw no reason why the electromagnetic laws should not be equally universal. The constancy of the speed of light and the universality of the laws of physics for all observers are cornerstones of special relativity.

Starting Points and Postulates

In developing special relativity, Einstein began by accepting what experiment and his own thinking showed to be the true behaviour of light, even when this contradicted classical physics or the usual perceptions about the world.

The fact that the speed of light is the same for all observers is inexplicable in ordinary terms. If a passenger in a train moving at 100 km per hour shoots an arrow in the train's direction of motion at 200 km per hour, a trackside observer would measure the speed of the arrow as the sum of the two speeds, or 300 km per hour. In analogy, if the train moves at the speed of light and a passenger shines a laser in the same direction, then common sense indicates that a trackside observer should see the light moving at the sum of the two speeds, or twice the speed of light (6×10^8 metres per second).

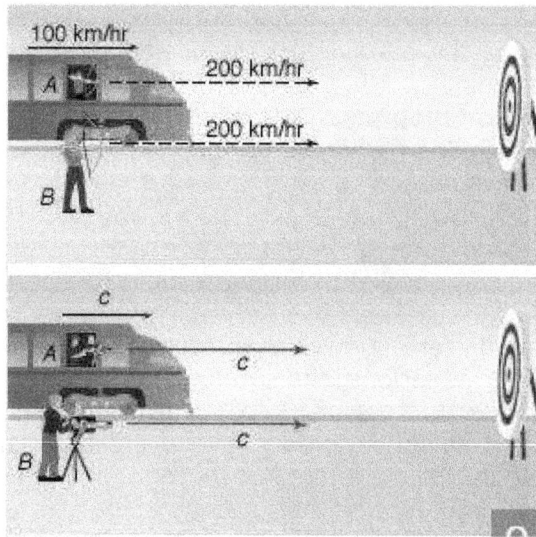

Invariance of the speed of lightArrows shot from a moving train (A) and from a stationary location (B) will arrive at a target at different velocities—in this case, 300 and 200 km/hr, respectively, because of the motion of the train. However, such commonsense addition of velocities does not apply to light.

Even for a train traveling at the speed of light, both laser beams, A and B, have the same velocity: c.

While such a law of addition of velocities is valid in classical mechanics, the Michelson-Morley experiment showed that light does not obey this law. This contradicts common sense; it implies, for instance, that both a train moving at the speed of light and a light beam emitted from the train arrive at a point farther along the track at the same instant.

Nevertheless, Einstein made the constancy of the speed of light for all observers a postulate of his new theory. As a second postulate, he required that the laws of physics have the same form for all observers. Then Einstein extended his postulates to their logical conclusions to form special relativity.

Consequences of the Postulates
Relativistic Space and Time

In order to make the speed of light constant, Einstein replaced absolute space and time with new definitions that depend on the state of motion of an observer. Einstein explained his approach by considering two observers and a train. One observer stands alongside a straight track; the other rides a train moving at constant speed along the track. Each views the world relative to his own surroundings. The fixed observer measures distance from a mark inscribed on the track and measures time with his watch; the train passenger measures distance from a mark inscribed on his railroad car and measures time with his own watch.

If time flows the same for both observers, as Newton believed, then the two frames of reference are reconciled by the relation: $x' = x - vt$. Here x is the distance to some specific event that happens along the track, as measured by the fixed observer; x' is the distance to the same event as measured by the moving observer; v is the speed of the train—that is, the speed of one observer relative to the other; and t is the time at which the event happens, the same for both observers. For example, suppose the train moves at 40 km per hour. One hour after it sets out, a tree 60 km from the train's starting point is struck by lightning. The fixed observer measures x as 60 km and t as one hour. The moving observer also measures t as one hour, and so, according to Newton's equation, he measures x' as 20 km.

Newtonian reference framesIsaac Newton reconciled different frames of reference with the equation $x' = x - vt$, where time (t) is assumed to be synchronous (that is, running at the same rate in both frames), x indicates the distance between an event and a stationary observer, x' indicates the distance between the same event and a moving observer, and v is the moving observer's velocity.

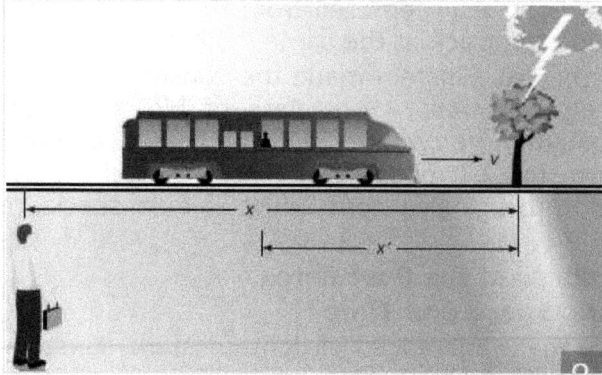

Newtonian reference frames Isaac Newton reconciled different frames of reference with the equation $x' = x - vt$, where time (t) is assumed to be synchronous (that is, running at the same rate in both frames), x indicates the distance between an event and a stationary observer, x' indicates the distance between the same event and a moving observer, and v is the moving observer's velocity.

This analysis seems obvious, but Einstein saw a subtlety hidden in its underlying assumptions—in particular, the issue of simultaneity. The two people do not actually observe the lightning strike at the same time. Even at the speed of light, the image of the strike takes time to reach each observer, and, since each is at a different distance from the event, the travel times differ. Taking this insight further, suppose lightning strikes two trees, one 60 km ahead of the fixed observer and the other 60 km behind, exactly as the moving observer passes the fixed observer. Each image travels the same distance to the fixed observer, and so he certainly sees the events simultaneously. The motion of the moving observer brings him closer to one event than the other, however, and he thus sees the events at different times.

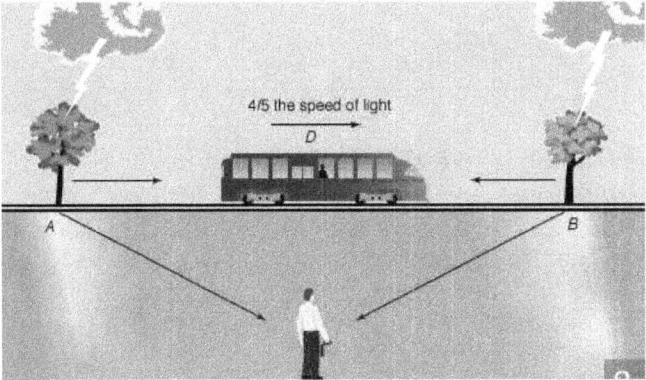

Simultaneous eventsSimultaneous events may
appear to coincide in time for one observer but not for another
because of differences in their spatial positions.

Einstein concluded that simultaneity is relative; events that are simultaneous for one observer may not be for another. This led him to the counterintuitive idea that time flows differently according to the state of motion and to the conclusion that distance is also relative. In the example, the train passenger and the fixed observer can each stretch a tape measure from back to front of a railroad car to find its length. The two ends of the tape must be placed in position at the same instant—that is, simultaneously—to obtain a true value. However, because the meaning of simultaneous is different for the two observers, they measure different lengths.

This reasoning led Einstein to new equations for time and space, called the Lorentz transformations, after the Dutch physicist Hendrik Lorentz, who first proposed them. They are:

$$x' = \frac{x - vt}{\sqrt{1 - \frac{v^2}{c^2}}} \quad \text{and} \quad t' = \frac{t - \frac{vx}{c^2}}{\sqrt{1 - \frac{v^2}{c^2}}}.$$

where t' is time as measured by the moving observer and c is the speed of light.

From these equations, Einstein derived a new relationship that replaces the classical law of addition of velocities,

$$u' = \frac{u + v}{1 + \frac{uv}{c^2}},$$

where u and u' are the speed of any moving object as seen by each observer and v is again the speed of one observer relative to the other. This relation guarantees Einstein's first postulate (that the speed of light is constant for all observers). In the case of the flashlight beam projected from a train moving at the speed of light, an observer on the train measures the speed of the beam as c. According to the equation above, so does the trackside observer, instead of the value 2c that classical physics predicts.

To make the speed of light constant, the theory requires that space and time change in a moving body, according to its speed, as seen by an outside observer. The body becomes shorter along its direction of motion; that is, its length contracts. Time intervals become longer, meaning that time runs more slowly in a moving body; that is, time dilates. In the train example, the person next to the track measures a shorter length for the train and a longer time interval for clocks on the train than does the train passenger. The relations describing these changes are relativistic length-timewhere L0 and T0, called proper length and proper time, respectively, are the values measured by an observer on the moving body, and L and T are the corresponding quantities as measured by a fixed observer.

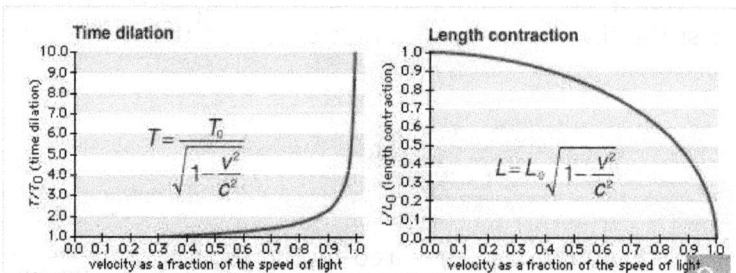

Length contraction and time dilationAs an object approaches the speed of light, an observer sees the object become shorter and its time interval become longer, relative to the length and time interval when the object is at rest.

The relativistic effects become large at speeds near that of light, although it is worth noting again that they appear only when an observer looks at a moving body. He never sees changes in space or time within his own reference frame (whether on a train or spacecraft), even at the speed of light. These effects do not appear in ordinary life, because the factor v2/c2 is minute at even the highest speeds attained by humans, so that Einstein's equations become virtually the same as the classical ones.

Relativistic Mass
Cosmic Speed Limit
To derive further results, Einstein combined his redefinitions of time and space with two powerful physical principles: conservation of energy and conservation of mass, which state that the total amount of each remains constant in a closed system. Einstein's second postulate ensured that these laws remained valid for all observers in the new theory, and he used them to derive the relativistic meanings of mass and energy.

One result is that the mass of a body increases with its speed. An observer on a moving body, such as a spacecraft, measures its so-called rest mass m_0, while a fixed observer measures its mass m as

$$m = \frac{m_0}{\sqrt{1 - \frac{v^2}{c^2}}},$$

which is greater than m_0. In fact, as the spacecraft's speed approaches that of light, the mass m approaches infinity. However, as the object's mass increases, so does the energy required to keep accelerating it; thus, it would take infinite energy to accelerate a material body to the speed of light. For this reason, no material object can reach the speed of light, which is the speed limit for the universe. (Light itself can attain this speed because the rest mass of a photon, the quantum particle of light, is zero.)

$$E = mc^2$$

Einstein's treatment of mass showed that the increased relativistic mass comes from the energy of motion of the body—that is, its kinetic energy E—divided by c2. This is the origin of the famous equation E = mc2, which expresses the fact that mass and energy are the same physical entity and can be changed into each other.

The Twin Paradox

The counterintuitive nature of Einstein's ideas makes them difficult to absorb and gives rise to situations that seem unfathomable. One well-known case is the twin paradox, a seeming anomaly in how special relativity describes time.

Suppose that one of two identical twin sisters flies off into space at nearly the speed of light. According to relativity, time runs more slowly on her spacecraft than on Earth; therefore, when she returns to Earth, she will be younger than her Earth-bound sister. But in relativity, what one observer sees as happening to a second one, the second one sees as happening to the first one. To the space-going sister, time moves more slowly on Earth than in her spacecraft; when she returns, her Earth-bound sister is the one who is younger. How can the space-going twin be both younger and older than her Earth-bound sister?

The answer is that the paradox is only apparent, for the situation is not appropriately treated by special relativity. To return to Earth, the spacecraft must change direction, which violates the condition of steady straight-line motion central to special relativity. A full treatment requires general relativity, which shows that there would be an asymmetrical change in time between the two sisters. Thus, the "paradox" does not cast doubt on how special relativity describes time, which has been confirmed by numerous experiments.

Four-Dimensional Space-Time

Special relativity is less definite than classical physics in that both the distance D and time interval T between two events depend on the observer. Einstein noted, however, that a particular combination of D and T, the quantity D2 − c2T2, has the same value for all observers.

The term cT in this invariant quantity elevates time to a kind of mathematical parity with space. Noting this, the German mathematical physicist Hermann Minkowski showed that the universe resembles a four-dimensional structure with coordinates x, y, z, and ct representing length, width, height, and time, respectively. Hence, the universe can be described as a four-dimensional space-time continuum, a central concept in general relativity.

Experimental Evidence For Special Relativity

Because relativistic changes are small at typical speeds for macroscopic objects, the confirmation of special relativity has relied on either the examination of subatomic bodies at high speeds or the measurement of small changes by sensitive instrumentation. For example, ultra-accurate clocks were placed on a variety of commercial airliners flying at one-millionth the speed of light. After two days of continuous flight, the time shown by the airborne clocks differed by fractions of a microsecond from that shown by a synchronized clock left on Earth, as predicted.

Larger effects are seen with elementary particles moving at speeds close to that of light. One such experiment involved muons, elementary particles created by cosmic rays in Earth's atmosphere at an altitude of about 9 km (30,000 feet). At 99.8 percent of the speed of light, the muons should reach sea level in 31 microseconds, but measurements showed that it took only 2 microseconds. The reason is that, relative to the moving muons, the distance of 9 km contracted to 0.58 km (1,900 feet). Similarly, a relativistic mass increase has been confirmed in measurements on fast-moving elementary particles, where the change is large.

Such results leave no doubt that special relativity correctly describes the universe, although the theory is difficult to accept at a visceral level. Some insight comes from Einstein's comment that in relativity the limiting speed of light plays the role of an infinite speed. At infinite speed, light would traverse any distance in zero time. Similarly, according to the relativistic equations, an observer riding a light wave would see lengths contract to zero and clocks stop ticking as the universe approached him at the speed of light. Effectively, relativity replaces an infinite speed limit with the finite value of 3×10^8 metres per second.

Roots of General Relativity

Because Isaac Newton's law of gravity served so well in explaining the behaviour of the solar system, the question arises why it was necessary to develop a new theory of gravity. The answer is that Newton's theory violates special relativity, for it requires an unspecified "action at a distance" through which any two objects—such as the Sun and Earth—instantaneously pull each other, no matter how far apart. However, instantaneous response would require the gravitational interaction to propagate at infinite speed, which is precluded by special relativity.

In practice, this is no great problem for describing our solar system, for Newton's law gives valid answers for objects moving slowly compared with light. Nevertheless, since Newton's theory cannot be conceptually reconciled with special relativity, Einstein turned to the development of general relativity as a new way to understand gravitation.

Principle of Equivalence

In order to begin building his theory, Einstein seized on an insight that came to him in 1907. As he explained in a lecture in 1922:

I was sitting on a chair in my patent office in Bern. Suddenly a thought struck me: If a man falls freely, he would not feel his weight. I was taken aback. This simple thought experiment made a deep impression on me. This led me to the theory of gravity.

Einstein was alluding to a curious fact known in Newton's time: no matter what the mass of an object, it falls toward Earth with the same acceleration (ignoring air resistance) of 9.8 metres per second squared. Newton explained this by postulating two types of mass: inertial mass, which resists motion and enters into his general laws of motion, and gravitational mass, which enters into his equation for the force of gravity. He showed that, if the two masses were equal, then all objects would fall with that same gravitational acceleration.

Einstein, however, realized something more profound. A person standing in an elevator with a broken cable feels weightless as the enclosure falls freely toward Earth. The reason is that both he and the elevator accelerate downward at the same rate and so fall at exactly same speed; hence, short of looking outside the elevator at his surroundings, he cannot determine that he is being pulled downward. In fact, there is no experiment he can do within a sealed falling elevator to determine that he is within a gravitational field. If he releases a ball from his hand, it will fall at the same rate, simply remaining where he releases it. And if he were to see the ball sink toward the floor, he could not tell if that was because he was at rest within a gravitational field that pulled the ball down or because a cable was yanking the elevator up so that its floor rose toward the ball.

Einstein expressed these ideas in his deceptively simple principle of equivalence, which is the basis of general relativity: on a local scale—meaning within a given system, without looking at other systems—it is impossible to distinguish between physical effects due to gravity and those due to acceleration.

In that case, continued Einstein's Gedankenexperiment, light must be affected by gravity. Imagine that the elevator has a hole bored straight through two opposite walls. When the elevator is at rest, a beam of light entering one hole travels in a straight line parallel to the floor and exits through the other hole. But if the elevator is accelerated upward, by the time the ray reaches the second hole, the opening has moved and is no longer aligned with the ray. As the passenger sees the light miss the second hole, he concludes that the ray has followed a curved path (in fact, a parabola).

If a light ray is bent in an accelerated system, then, according to the principle of equivalence, light should also be bent by gravity, contradicting the everyday expectation that light will travel in a straight line (unless it passes from one medium to another). If its path is curved by gravity, that must mean that "straight line" has a different meaning near a massive gravitational body such as a star than it does in empty space. This was a hint that gravity should be treated as a geometric phenomenon.

Curved Space-Time And Geometric Gravitation

The singular feature of Einstein's view of gravity is its geometric nature. Whereas Newton thought that gravity was a force, Einstein showed that gravity arises from the shape of space-time. While this is difficult to visualize, there is an analogy that provides some insight—although it is only a guide, not a definitive statement of the theory.

The analogy begins by considering space-time as a rubber sheet that can be deformed. In any region distant from massive cosmic objects such as stars, space-time is uncurved—that is, the rubber sheet is absolutely flat. If one were to probe space-time in that region by sending out a ray of light or a test body, both the ray and the body would travel in perfectly straight lines, like a child's marble rolling across the rubber sheet.

However, the presence of a massive body curves space-time, as if a bowling ball were placed on the rubber sheet to create a cuplike depression. In the analogy, a marble placed near the depression rolls down the slope toward the bowling ball as if pulled by a force. In addition, if the marble is given a sideways push, it will describe an orbit around the bowling ball, as if a steady pull toward the ball is swinging the marble into a closed path.

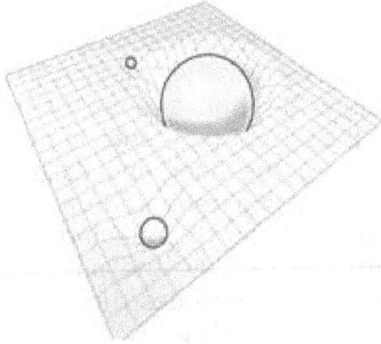

Curved space-timeThe four dimensional space-time continuum itself is distorted in the vicinity of any mass, with the amount of distortion depending on the mass and the distance from the mass. Thus, relativity accounts for Newton's inverse square law of gravity through geometry and thereby does away with the need for any mysterious "action at a distance."

In this way, the curvature of space-time near a star defines the shortest natural paths, or geodesics—much as the shortest path between any two points on Earth is not a straight line, which cannot be constructed on that curved surface, but the arc of a great circle route. In Einstein's theory, space-time geodesics define the deflection of light and the orbits of planets. As the American theoretical physicist John Wheeler put it, matter tells space-time how to curve, and space-time tells matter how to move.

The Mathematics of General Relativity

The rubber sheet analogy helps with visualization of space-time, but Einstein himself developed a complete quantitative theory that describes space-time through highly abstract mathematics. General relativity is expressed in a set of interlinked differential equations that define how the shape of space-time depends on the amount of matter (or, equivalently, energy) in the region. The solution of these so-called field equations can yield answers to different physical situations, including the behaviour of individual bodies and of the entire universe.

Cosmological Solutions

Einstein immediately understood that the field equations could describe the entire cosmos. In 1917 he modified the original version of his equations by adding what he called the "cosmological term." This represented a force that acted to make the universe expand, thus counteracting gravity, which tends to make the universe contract. The result was a static universe, in accordance with the best knowledge of the time.

In 1922, however, the Soviet mathematician Aleksandr Aleksandrovich Friedmann showed that the field equations predict a dynamic universe, which can either expand forever or go through cycles of alternating expansion and contraction. Einstein came to agree with this result and abandoned his cosmological term. Later work, notably pioneering measurements by the American astronomer Edwin Hubble and the development of the big-bang model, has confirmed and amplified the concept of an expanding universe.

Black Holes

In 1916 the German astronomer Karl Schwarzschild used the field equations to calculate the gravitational effect of a single spherical body such as a star. If the mass is neither very large nor highly concentrated, the resulting calculation will be the same as that given by Newton's theory of gravity. Thus, Newton's theory is not incorrect; rather, it constitutes a valid approximation to general relativity under certain conditions.

Schwarzschild also described a new effect. If the mass is concentrated in a vanishingly small volume—a singularity—gravity will become so strong that nothing pulled into the surrounding region can ever leave. Even light cannot escape. In the rubber sheet analogy, it as if a tiny massive object creates a depression so steep that nothing can escape it. In recognition that this severe space-time distortion would be invisible—because it would absorb light and never emit any—it was dubbed a black hole.

In quantitative terms, Schwarzschild's result defines a sphere that is centred at the singularity and whose radius depends on the density of the enclosed mass. Events within the sphere are forever isolated from the remainder of the universe; for this reason, the Schwarzschild radius is called the event horizon.

Black Holes And Wormholes

No human technology could compact matter sufficiently to make black holes, but they may occur as final steps in the life cycle of stars. After millions or billions of years, a star uses up all of its hydrogen and other elements that produce energy through nuclear fusion. With its nuclear furnace banked, the star no longer maintains an internal pressure to expand, and gravity is left unopposed to pull inward and compress the star. For stars above a certain mass, this gravitational collapse will in principle produce a black hole containing several times the mass of the Sun. In other cases, the gravitational collapse of huge dust clouds may create supermassive black holes containing millions or billions of solar masses.

Astrophysicists have found several cosmic objects that appear to contain a dense concentration of mass in a small volume. These strong candidates for black holes include one at the centre of the Milky Way Galaxy and certain binary stars that emit X-rays as they orbit each other. However, the definitive signature of a black hole, the event horizon, has not been observed.

The theory of black holes has led to another predicted entity, a wormhole. This is a solution of the field equations that resembles a tunnel between two black holes or other points in space-time. Such a tunnel would provide a shortcut between its end points. In analogy, consider an ant walking across a flat sheet of paper from point A to point B. If the paper is curved through the third dimension, so that A and B overlap, the ant can step directly from one point to the other, thus avoiding a long trek.

The possibility of short-circuiting the enormous distances between stars makes wormholes attractive for space travel. Because the tunnel links moments in time as well as locations in space, it also has been argued that a wormhole would allow travel into the past. However, wormholes are intrinsically unstable. While exotic stabilization schemes have been proposed, there is as yet no evidence that these can work or indeed that wormholes exist.

Experimental Evidence For General Relativity

Soon after the theory of general relativity was published in 1915, the English astronomer Arthur Eddington considered Einstein's prediction that light rays are bent near a massive body, and he realized that it could be verified by carefully comparing star positions in images of the Sun taken during a solar eclipse with images of the same region of space taken when the Sun was in a different portion of the sky. Verification was delayed by World War I, but in 1919 an excellent opportunity presented itself with an especially long total solar eclipse, in the vicinity of the bright Hyades star cluster, that was visible from northern Brazil to the African coast. Eddington led one expedition to Príncipe, an island off the African coast, and Andrew Crommelin of the Royal Greenwich Observatory led a second expedition to Sobral, Brazil. After carefully comparing photographs from both expeditions with reference photographs of the Hyades, Eddington declared that the starlight had been deflected about 1.75 seconds of arc, as predicted by general relativity. (The same effect produces gravitational lensing, where a massive cosmic object focuses light from another object beyond it to produce a distorted or magnified image. The astronomical discovery of gravitational lenses in 1979 gave additional support for general relativity.)

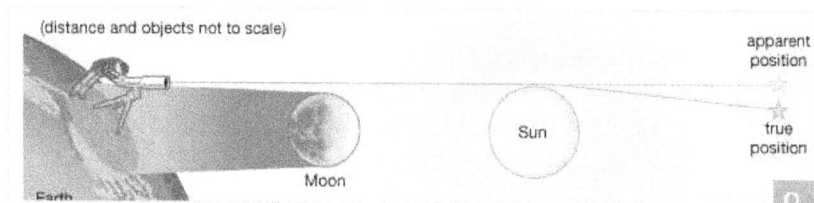

(distance and objects not to scale)

apparent position

true position

Sun

Moon

Earth

Experimental evidence for general relativityIn 1919, observation of a solar eclipse confirmed Einstein's prediction that light is bent in the presence of mass. This experimental support for his general theory of relativity garnered him instant worldwide acclaim.

Further evidence came from the planet Mercury. In the 19th century, it was found that Mercury does not return to exactly the same spot every time it completes its elliptical orbit.

Instead, the ellipse rotates slowly in space, so that on each orbit the perihelion—the point of closest approach to the Sun—moves to a slightly different angle. Newton's law of gravity could not explain this perihelion shift, but general relativity gave the correct orbit.

Another confirmed prediction of general relativity is that time dilates in a gravitational field, meaning that clocks run slower as they approach the mass that is producing the field. This has been measured directly and also through the gravitational redshift of light. Time dilation causes light to vibrate at a lower frequency within a gravitational field; thus, the light is shifted toward a longer wavelength—that is, toward the red. Other measurements have verified the equivalence principle by showing that inertial and gravitational mass are precisely the same.

The most striking prediction of general relativity is that of gravitational waves. Electromagnetic waves are caused by accelerated electrical charges and are detected when they put other charges into motion. Similarly, gravitational waves would be caused by masses in motion and are detected when they initiate motion in other masses. However, gravity is very weak compared with electromagnetism. Only a huge cosmic event, such as the collision of two stars, can generate detectable gravitational waves. Efforts to sense gravitational waves began in the 1960s, and such waves were first detected in 2015 when LIGO observed two black holes 1.3 million light-years away spiralling into each other.

Applications of Relativistic Ideas
Although relativistic effects are negligible in ordinary life, relativistic ideas appear in a range of areas from fundamental science to civilian and military technology.

Elementary Particles
The relationship $E = mc^2$ is essential in the study of subatomic particles. It determines the energy required to create particles or to convert one type into another and the energy released when a particle is annihilated. For example, two photons, each of energy E, can collide to form two particles, each with mass $m = E/c^2$. This pair-production process is one step in the early evolution of the universe, as described in the big-bang model.

Particle accelerators

Knowledge of elementary particles comes primarily from particle accelerators. These machines raise subatomic particles, usually electrons or protons, to nearly the speed of light. When these energetic bullets smash into selected targets, they elucidate how subatomic particles interact and often produce new species of elementary particles.

Particle accelerators could not be properly designed without special relativity. In the type called an electron synchrotron, for instance, electrons gain energy as they traverse a huge circular raceway. At barely below the speed of light, their mass is thousands of times larger than their rest mass. As a result, the magnetic field used to hold the electrons in circular orbits must be thousands of times stronger than if the mass did not change.

Fission and Fusion: Bombs and Stellar Processes

Energy is released in two kinds of nuclear processes. In nuclear fission a heavy nucleus, such as uranium, splits into two lighter nuclei; in nuclear fusion two light nuclei combine into a heavier one. In each process the total final mass is less than the starting mass. The difference appears as energy according to the relation $E = \Delta mc^2$, where Δm is the mass deficit.

Fission is used in atomic bombs and in reactors that produce power for civilian and military applications. The fusion of hydrogen into helium is the energy source in stars and provides the power of a hydrogen bomb. Efforts are now under way to develop controllable hydrogen fusion as a clean, abundant power source.

The Global Positioning System

The global positioning system (GPS) depends on relativistic principles. A GPS receiver determines its location on Earth's surface by processing radio signals from four or more satellites. The distance to each satellite is calculated as the product of the speed of light and the time lag between transmission and reception of the signal. However, Earth's gravitational field and the motion of the satellites cause time-dilation effects, and Earth's rotation also has relativistic implications. Hence, GPS technology includes relativistic corrections that enable positions to be calculated to within several centimetres.

Cosmology

Cosmology, the study of the structure and origin of the universe, is intimately connected with gravity, which determines the macroscopic behaviour of all matter. General relativity has played a role in cosmology since the early calculations of Einstein and Friedmann. Since then, the theory has provided a framework for accommodating observational results, such as Hubble's discovery of the expanding universe in 1929, as well as the big-bang model, which is the generally accepted explanation of the origin of the universe.

The latest solutions of Einstein's field equations depend on specific parameters that characterize the fate and shape of the universe. One is Hubble's constant, which defines how rapidly the universe is expanding; the other is the density of matter in the universe, which determines the strength of gravity. Below a certain critical density, gravity would be weak enough that the universe would expand forever, so that space would be unlimited. Above that value, gravity would be strong enough to make the universe shrink back to its original minute size after a finite period of expansion, a process called the "big crunch." In this case, space would be limited or bounded like the surface of a sphere. Current efforts in observational cosmology focus on measuring the most accurate possible values of Hubble's constant and of critical density.

Relativity, Quantum Theory, And Unified Theories

Cosmic behaviour on the biggest scale is described by general relativity. Behaviour on the subatomic scale is described by quantum mechanics, which began with the work of the German physicist Max Planck in 1900 and treats energy and other physical quantities in discrete units called quanta. A central goal of physics has been to combine relativity theory and quantum theory into an overarching "theory of everything" describing all physical phenomena. Quantum theory explains electromagnetism and the strong and weak forces, but a quantum description of the remaining fundamental force of gravity has not been achieved.

After Einstein developed relativity, he unsuccessfully sought a so-called unified field theory with a space-time geometry that would encompass all the fundamental forces. Other theorists have attempted to merge general relativity with quantum theory, but the two approaches treat forces in fundamentally different ways. In quantum theory, forces arise from the interchange of certain elementary particles, not from the shape of space-time. Furthermore, quantum effects are thought to cause a serious distortion of space-time at an extremely small scale called the Planck length, which is much smaller than the size of elementary particles. This suggests that quantum gravity cannot be understood without treating space-time at unheard-of scales.

Although the connection between general relativity and quantum mechanics remains elusive, some progress has been made toward a fully unified theory. In the 1960s, the electroweak theory provided partial unification, showing a common basis for electromagnetism and the weak force within quantum theory. Recent research suggests that superstring theory, in which elementary particles are represented not as mathematical points but as extremely small strings vibrating in 10 or more dimensions, shows promise for supporting complete unification, including gravitation. However, until confirmed by experimental results, superstring theory will remain an untested hypothesis.

Intellectual And Cultural Impact of Relativity
Reactions in General Culture

The impact of relativity has not been limited to science. Special relativity arrived on the scene at the beginning of the 20th century, and general relativity became widely known after World War I—eras when a new sensibility of "modernism" was becoming defined in art and literature. In addition, the confirmation of general relativity provided by the solar eclipse of 1919 received wide publicity. Einstein's 1921 Nobel Prize for Physics (awarded for his work on the photon nature of light), as well as the popular perception that relativity was so complex that few could grasp it, quickly turned Einstein and his theories into cultural icons.

The ideas of relativity were widely applied—and misapplied—soon after their advent. Some thinkers interpreted the theory as meaning simply that all things are relative, and they employed this concept in arenas distant from physics. The Spanish humanist philosopher and essayist José Ortega y Gasset, for instance, wrote in The Modern Theme (1923),

The theory of Einstein is a marvelous proof of the harmonious multiplicity of all possible points of view. If the idea is extended to morals and aesthetics, we shall come to experience history and life in a new way.

The revolutionary aspect of Einstein's thought was also seized upon, as by the American art critic Thomas Craven, who in 1921 compared the break between classical and modern art to the break between Newtonian and Einsteinian ideas about space and time.

Some saw specific relations between relativity and art arising from the idea of a four-dimensional space-time continuum. In the 19th century, developments in geometry led to popular interest in a fourth spatial dimension, imagined as somehow lying at right angles to all three of the ordinary dimensions of length, width, and height. Edwin Abbott's Flatland (1884) was the first popular presentation of these ideas. Other works of fantasy that followed spoke of the fourth dimension as an arena apart from ordinary existence.

Einstein's four-dimensional universe, with three spatial dimensions and one of time, is conceptually different from four spatial dimensions. But the two kinds of four-dimensional world became conflated in interpreting the new art of the 20th century. Early Cubist works by Pablo Picasso that simultaneously portrayed all sides of their subjects became connected with the idea of higher dimensions in space, which some writers attempted to relate to relativity. In 1949, for example, the art historian Paul LaPorte wrote that "the new pictorial idiom created by [C]ubism is most satisfactorily explained by applying to it the concept of the space-time continuum." Einstein specifically rejected this view, saying, "This new artistic 'language' has nothing in common with the Theory of Relativity." Nevertheless, some artists explicitly explored Einstein's ideas. In the new Soviet Union of the 1920s, for example, the poet and illustrator Vladimir Mayakovsky, a founder of the artistic movement called Russian Futurism, or Suprematism, hired an expert to explain relativity to him.

The widespread general interest in relativity was reflected in the number of books written to elucidate the subject for nonexperts. Einstein's popular exposition of special and general relativity appeared almost immediately in 1916 . Other scientists, such as the Russian mathematician Aleksandr Friedmann and the British astronomer Arthur Eddington, wrote popular books on the subjects in the 1920s. Such books continued to appear decades later.

When relativity was first announced, the public was typically awestruck by its complexity, a justified response to the intricate mathematics of general relativity. But the abstract, nonvisceral nature of the theory also generated reactions against its apparent violation of common sense. These reactions included a political undertone; in some quarters, it was considered undemocratic to present or support a theory that could not be immediately understood by the common person.

In contemporary usage, general culture has accepted the ideas of relativity—the impossibility of faster-than-light travel, E = mc2, time dilation and the twin paradox, the expanding universe, and black holes and wormholes—to the point where they are immediately recognized in the media and provide plot devices for works of science fiction. Some of these ideas have gained meaning beyond their strictly scientific ones; in the business world, for instance, "black hole" can mean an unrecoverable financial drain.

Philosophical Considerations

In 1925 the British philosopher Bertrand Russell, in his ABC of Relativity, suggested that Einstein's work would lead to new philosophical concepts. Relativity has indeed had a great effect on philosophy, illuminating some issues that go back to the ancient Greeks. The idea of the ether, invoked in the late 19th century to carry light waves, harks back to Aristotle. He divided the world into earth, air, fire, and water, with the ether (aether) as the fifth element representing the pure celestial sphere. The Michelson-Morley experiment and relativity eliminated the last vestiges of this idea.

Relativity also changed the meaning of geometry as it was developed in Euclid's Elements (c. 300 BCE). Euclid's system relied on the axiom "a straight line is the shortest distance between two points," among others that seemed self-evidently true. Straight lines also played a special role in Euclid's Optics as the paths followed by light rays. To philosophers such as the German Immanuel Kant, Euclid's straight-line axiom represented a deep level of truth. But general relativity makes it possible scientifically to examine space like any other physical quantity—that is, to investigate Euclid's premises. It is now known that space-time is curved near stars; no straight lines exist there, and light follows curved geodesics. Like Newton's law of gravity, Euclid's geometry correctly describes reality under certain conditions, but its axioms are not absolutely fundamental and universal, for the cosmos includes non-Euclidean geometries as well.

Considering its scientific breadth, its recasting of people's view of reality, its ability to describe the entire universe, and its influence outside science, Einstein's relativity stands among the most significant and influential of scientific theories.

*****

Quantum Computing

The field of quantum computing was pioneered in 1985 by Daved Deutsch. Building upon a suggestion by Feynman and the work of other scientists, he generalized the concept of the Turing Machine as postulated by Turing. He invoked quantum mechanics, our most accurate description of reality till now—realizing that the Turing Machine had the implicit assumption of classical mechanics—and reworked the Turing Machine. The concept behind quantum computing is simple. Use quantum mechanical systems with the full use of their quantum mechanical properties to do computations. Why is this useful? As was shown over the years, quantum mechanical dynamics can be used to compute the answers to certain problems much faster than any classical system or computer can

For instance numbers can be factored in polynomial time by quantum computers, compared to the exponential time of classical ones. This has brought about a fundamental revolution in the field of complexity theory and cryptography, among others.

The problem, however, is that no large scale quantum computer has ever been built. The largest such system was just powerful enough to factor 15 into 3 × 5 with high probability! Needless to say, we are a long way off from the construction of sufficiently powerful quantum computers. By sufficiently powerful, I mean powerful enough to solve problems that current classical computers can do and preferably much more. At the heart of quantum computing lies the concept of quantum superposition. A classical system can only be in one state at a time. A quantum system can be in a superposition of multiple states at a time. Hence, the idea is to perform computation in parallel. Though it's much trickier than the parrallel processing implemented in networked classical computers today.There are two reasons for this. First, one can set the initial state of a quantum computers into a superposition of states. But when you evolve that state—or closer to the language of computing, you perform computation n it—the different states do not remain separate. If two different states evolve into the same final state, there is no way to distinguish how much amplitude is due to each of those initial states. This puts a huge limit on the type of quantum algorithms we an design, and explains why we can't just import classical network computing algorithms to the field of quantum computing.The second reason is that of measurement. This is, in fact, even a more fundamental restriction on what sort of computations we can do. Recall that whatever the state of a quantum system, a measurement on it only gives us one of the possible eigenvalues. Therefore, even if we evolve a system into a superposition of final states, we only get to find out one of them. Repeating the computation or experiment over and over again is of not much use either, because it defeats the purpose of quantum computers solving problems faster. Nevertheless, clever algorithms have been designed that give the answer to a problem in just one measurement. Though this is not the only useful construction of quantum algorithms.Nevertheless, quantum computing holds enormous promise. It has changed fundamentally how we think about physics and computer science. Moreover, it promises great technological innovations. In the following sections, I present an introductory level explanation of quantum computing.

Modelling Quantum Computers
There are two fundamental ways that we can look at the theoretical construction of quantum computers.

Turing Model
The quantum computer is fundamentally the same theoretical object as a classical computer as far as computation is concerned. This means that a quantum computer can compute nothing a classical computer cannot and vice versa. Since, a classical computer is equivalent to a Universal Turing Machine, so is a quantum computer. In other words, quantum computers may make certain problems tractable i.e. make it possible to compute them in polynomial time compared to relatively slower algorithms for classical computers. The Turing model might be great for a core theoretical understanding of a computer. However, for practical implementation a different model is needed.

The Circuit Model of Quantum Computation

The circuit model is essentially similar to the circuit model for classical computers. It allows us to (relatively) easily visualize the state of the system at any point in the computation given a definite input. An example of such a circuit model is given in 1. This is the circuit for the quantum Fourier transform, which we will encounter many times in these notes.

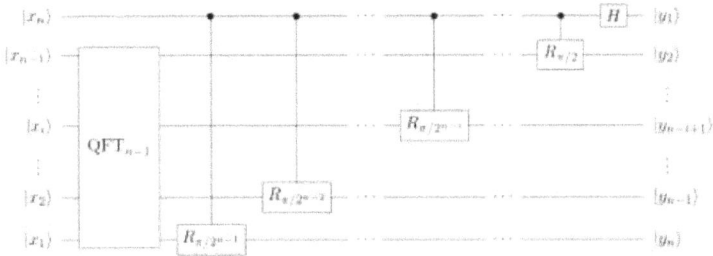

Figure 1: The circuit for the Quantum Fourier Transform

It shows all the major features of the quantum circuit model: quantum wires, quantum gates and qubits. We will discuss all these features in greater detail.

Qubits The basic information resource in quantum computation is the qubit, which is derived from "quantum bit". Essentially, all the information being that is manipulated during the course of a quantum computation is stored in registers of qubits. A single qubit is a two state quantum system. A register is a set of qubits. We know from quantum mechanics that a two state system in quantum mechanics can be in any superposition of the two basis states. Conventionally, the two basis states in quantum computating literature are represented by |0> and |1>. This allows us to write the state of a single qubit as

$$|\psi\rangle = \alpha |0\rangle + \beta |1\rangle$$

α and β are in general complex, and due to the normalization condition of state vectors $\alpha * \alpha + \beta * \beta = 1$. Here we have

$$|0\rangle = \begin{bmatrix} 1 \\ 0 \end{bmatrix}$$

And

$$|1\rangle = \begin{bmatrix} 0 \\ 1 \end{bmatrix}$$

This means that :

$$|\psi\rangle = \alpha \begin{bmatrix} 1 \\ 0 \end{bmatrix} + \beta \begin{bmatrix} 0 \\ 1 \end{bmatrix}$$

In other words, generic quantum states can be represented by column matrices. To represent a register of qubits we use the following scheme. For a two qubit system, the possible states are |00>, |01>, |10>, |11>. This means that an arbitrary state fo a two qubit system can be represented by

$$|\psi\rangle = \alpha_{00} |00\rangle + \alpha_{01} |01\rangle + \alpha_{10} |10\rangle + \alpha_{11} |11\rangle$$

In column matrix formulation, the basis states are

$$|00\rangle = \begin{bmatrix} 1 \\ 0 \\ 0 \\ 0 \end{bmatrix} \qquad |01\rangle = \begin{bmatrix} 0 \\ 1 \\ 0 \\ 0 \end{bmatrix}$$

$$|10\rangle = \begin{bmatrix} 0 \\ 0 \\ 1 \\ 0 \end{bmatrix} \qquad |11\rangle = \begin{bmatrix} 0 \\ 0 \\ 0 \\ 1 \end{bmatrix}$$

This simple scheme can be generalized for a N qubit system to give us

$$|\psi\rangle = \sum_{i,j,\ldots,m \in \{0,1\}^N} \alpha_{ij\ldots m} |ij \ldots m\rangle$$

Quantum Gates

To perform operations on qubits we require quantum gates. Quantum gates are essentially evolutions of quantum states. In the circuit model they can be treated as blackboxes, meaning

300

that for the time being we don't care how they are physically implemented. All we care about the inputs they take and the corresponding outputs. Quantum gates have the property that they have an equivalent number of inputs and outputs. This stems specifically from the fact that quantum mechanics is a theory that obeys time symmetery. A quantum mechanical system that is evolved from a given initial state to a final state using a definite series of operations can be evolved backwards from the final state to the initial state using the inverse of the series of operations on it.

A quantum computer is also a reversible machine. A gate operating on an input to turn it into an output will not be able to do the reverse if the number of outputs are smaller than the number of inputs because information would be lost in the process. Hence, the equivalence of the number of inputs and output. We can divide quantum gates intocategories based on the number of inputs/outputs. To represent the operation of particular gates, two schemes are used. The first one is the more abstract one. It involves giving the operation of the gate on the basis states. Since, any quantum state can be decomposed as a linear combination of the basis states, giving the operation on the basis states is sufficient to describe the operation of the gate on any state. We will use this scheme later on for large qubit systems. The second representation involves matrices. Since, the state of a qubit register can be given by kets or column matrices, operations on them can be represented by square matrices.Multiply an input state column matrix with the matrix for the quantum gate, we obtain the output state. This scheme makes conceptualizing quantum computing easier. However, it is not practical for large systems, where the matrices would be too large to yield on paper or the computer screen.

Single Qubit Gates
The simplest quantum gates are single qubits gates. We look at a couple of such gates.

Identity Gate Of these the simplest gate is the identity gate. This identically outputs the input. It's typically represented by I. In the first scheme, we have

$$I\,|0\rangle = |0\rangle$$
$$I\,|1\rangle = |1\rangle.$$

This completely specifies the operation of the identity gate. Given the generic state

$|\psi\rangle = \alpha\,|0\rangle + \beta\,|1\rangle$, we can input it into the identity gate to get out

$$I\,|\psi\rangle = I(\alpha\,|0\rangle + \beta\,|1\rangle)$$
$$= \alpha\,|0\rangle + \beta\,|1\rangle.$$

In the matrix representation, the I is given by the identity matrix

$$I = \begin{bmatrix} 1 & 0 \\ 0 & 1 \end{bmatrix}.$$

Multiply in any state with I will just return the original state.
NOT Gate The NOT gate switches between the two basis states

$$X\,|0\rangle = |1\rangle$$
$$X\,|1\rangle = |0\rangle$$

The reason for representing the NOT gate by X is because it's matrix is given by

$$X = \begin{bmatrix} 0 & 1 \\ 1 & 0 \end{bmatrix}.$$

This is just one of the Pauli matrices, one which is usually represented by X. We can check it's operation on the two basis states

$$\begin{bmatrix} 0 & 1 \\ 1 & 0 \end{bmatrix} \begin{bmatrix} 1 \\ 0 \end{bmatrix} = \begin{bmatrix} 0 \\ 1 \end{bmatrix}$$

And

$$\begin{bmatrix} 0 & 1 \\ 1 & 0 \end{bmatrix} \begin{bmatrix} 0 \\ 1 \end{bmatrix} = \begin{bmatrix} 1 \\ 0 \end{bmatrix}$$

The other two Pauli matrices are gates in their own right, though their operation is slightly more complicated. We have

$$Y = \begin{bmatrix} 0 & -i \\ i & 0 \end{bmatrix}$$

And

$$Z = \begin{bmatrix} 1 & 0 \\ 0 & -1 \end{bmatrix}.$$

Hadamard Gate This is a special type of gate which takes a single basis state to a superposition of states

$$\mathbf{H}\,|0\rangle = \frac{1}{\sqrt{2}}(|0\rangle + |1\rangle)$$

$$\mathbf{H}\,|1\rangle = \frac{1}{\sqrt{2}}(|0\rangle - |1\rangle)$$

It's matrix is given by

$$\frac{1}{\sqrt{2}}\begin{bmatrix} 1 & 1 \\ 1 & -1 \end{bmatrix}.$$

Notable about this gate is that it is self inverse, i.e. $H = H^{-1}$.

This means that

$$\mathbf{H}\left(\frac{1}{\sqrt{2}}(|0\rangle + |1\rangle)\right) = |0\rangle,$$

And

$$\mathbf{H}\left(\frac{1}{\sqrt{2}}(|0\rangle - |1\rangle)\right) = |1\rangle.$$

Phase Gate This applies a phase difference to the qubit to only the |1i state, while
leaving the |0> state unchanged.

$$\mathbf{R}\,|0\rangle = |1\rangle$$

$$\mathbf{R}\,|1\rangle = e^{i\phi}\,|1\rangle$$

It's matrix is given by

$$\begin{bmatrix} 1 & 0 \\ 0 & e^{i\phi} \end{bmatrix}$$

Two Qubit and Higher Dimensional Gates
Two (and higher) qubit gates are essentially are of two types. One type is the analog of classical gates, where the output depends on all the inputs. The other type is what are termed 'controlled' gates. These have the feature that one of the input/output pair— called the target qubit—has an action done on it if and only if the other input/output pairs—called the control qubits—have a certain value. The control qubits are unaffected by the gate in either case. We look at a couple of such cases.

C-NOT Gate Here the C-NOT stands for 'Controlled NOT'. It applies the NOT gate to the target qubit if and only if the control qubit is |1i. It's matrix representation is given by

$$\begin{bmatrix} 1 & 0 & 0 & 0 \\ 0 & 1 & 0 & 0 \\ 0 & 0 & 0 & 1 \\ 0 & 0 & 1 & 0 \end{bmatrix}$$

This gate can equivalently be seen as one which XORs the value of the second qubit with the value of the first. Essentially, under the action of this gate,

$$|x\rangle |y\rangle \rightarrow |x\rangle |x \oplus y\rangle .$$

Controlled Phase Shift On the same lines as the C-NOT, this applies a phase shift to the target qubit if the controlled qubit has state |1>.

$$\begin{bmatrix} 1 & 0 & 0 & 0 \\ 0 & 1 & 0 & 0 \\ 0 & 0 & 0 & 1 \\ 0 & 0 & 0 & e^{i\phi} \end{bmatrix}$$

Hadamard Gate The single qubit Hadamard Gate can be generalized to a n-qubit gate, where the single qubit Hadamard Gate is applied separately and individually to all the qubits.

$$\mathbf{H} |x_0\rangle |x_1\rangle |x_2\rangle \cdots |x_{n-1}\rangle \rightarrow (\mathbf{H} |x_0\rangle)(\mathbf{H} |x_1\rangle)(\mathbf{H} |x_2\rangle) \cdots (\mathbf{H} |x_{n-1}\rangle)$$

Function Evaluations

Gates can be stacked together in a quantum network to build up more complicated gates.Some of these gates can become the incarnation of functions. For instance, the C-NOT gate is single gate that allows us to XOR the two input values.A more complicated example is shown in Fig. 2. This shows a quantum adder, that building on the same principles as the XOR, allows us to fully add two numbers.Many times, in the course of quantum computation, we are in need of such quantum networks. Typically, we are not interested in how a particular function is implemented. Sometimes, we are interested in generic functions of only a certain type. In those case, we abstract away all details, and treat these functions as blackboxes. In other words,these functions can be expressed as quantum gates, which take in a certain number of inputs and transform them in a definite way to an output.

Implementing Quantum Computers

Implementing practical quantum computation is no mean feat. It involves technological feats that have not been imvented yet. However, the great thing is that workers in the field have decided on a sufficiently general framework for the theoretical construction of quantum computers that it allows wide scope for the implementation of such systems. This has driven innovations in various directions. However, as yet no practical model has been decided as the preferred way of implementing quantum computers.

DiVincenzo Criteria

In 1997 David P. DiVincenzo set up five essential criteria that any quantum computer must fulfill, so as to make the connection with the theoretical circuit model complete.

1. Hilbert space control. The Hilbert space we work with must be precisely definablei.e. we control the exact space we are working with. It must also be possible to scale the system. This means that more qubits can be added by a simple tensor product scheme.

2. State preparation. It must be possible to prepare qubits in some some simple known state from which computation can proceed.

3. Low decoherence. It must be possible to isolate our quantum computer or quantum mechanical system from the environment so that decoherence is low. There is no known analog of Shannon's noisy channel coding theorem. Consequently, our knowledge of acceptable decoherence level is dependant on the most effecient error correcting codes for quantum computation known.

4. Implementation of quantum gates. Quantum gates form a crucial part of quantum computers. Basic quantum gates are simply"controlled sequences of precisely defined unitary transformation". They must be implemented with high precision and it must be possible to target specific groups of qubits to perform these translations on.

5. State specific quantum measurements. At the end of any quantum computations,and sometimes in between, measurements are made to read out the state of the system. It must be possible to be able to perform qubit specific measurements as well as on group of subsystems. Moreover, it must be possible to perform such measurements in whatever basis we choose.

Quantum Entanglement & Quantum Tunneling
Quantum Entanglement

In the day-to-day world that is well described by classical physics, we often observe correlations. Imagine you are observing a bank robbery. The bank robber is pointing a gun at the terrified teller. By looking at the teller you can tell whether the gun has gone off or not. If the teller is alive and unharmed, you can be sure the gun has not fired. If the teller is lying dead of a gun-shot wound on the floor, you know that the gun has fired.

This is elementary detective work. On the other hand, by examining the gun to see whether it has fired, you can find out whether the teller is alive or dead. We could say that there is a direct correlation between the state of the gun and the state of the teller. 'Gun fired' means 'teller dead', and 'gun not-fired' means 'teller alive'. We assume that the robber only shoots to kill and he never misses.

In the world of microscopic objects described by quantum mechanics, things are not always so simple. Imagine an atom which might undergo a radioactive decay in a certain time, or it might not. We might expect that with respect to the decay, there are only two possible states here: 'decayed', and 'not decayed', just as we had two states, 'fired' and 'not fired' for the gun or 'alive' and 'dead' for the teller. However, in the quantum mechanical world, it is also possible for the atom to be in a combined state 'decayed-not decayed' in which it is neither one nor the other but somewhere in between. This is called a 'superposition' of the two states, and is not something we normally expect of classical objects like guns or tellers. Two atoms may be correlated so that if the first has decayed, the second will also have decayed, and if the first atom has not decayed, neither has the second. This is a 100% correlation. But the quantum mechanical atoms may also be correlated so that if the first is in the superposition 'decayed-not decayed', the second will be also. Quantum mechanically there are more correlations between the atoms than we would expect classically. This kind of quantum 'super-correlation' is called 'entanglement'.

Entanglement was in fact originally named in German, 'Verschrankung', by Schrodinger, who was one of the first people to realise how strange it was. Imagine it is not the robber but the atom which determines whether the gun fires. If the atom decays it sets off a hair trigger which fires the gun. If it doesn't decay, the gun doesn't fire. But what does it mean if the atom is in the superposition state 'decayed-not decayed'? Then can it be correlated to the gun in a superposition state 'fired-not fired'? And what about the poor teller, who is now dead and alive at the same time? Schrodinger was worried by a similar situation where the victim of the quantum entanglement was a cat in a box where the decaying atom could trigger the release of a lethal chemical. The problem is that in the everyday world we are not used to seeing anything like a 'dead-live' cat, or a 'dead-live' teller, but in principle, if we expect quantum mechanics to be a complete theory describing every level of our experience, such strange states should be possible. Where does the strange quantum world stop and the ordinary classical world begin? These are problems which have now been debated for decades, and a number of different 'interpretations' of the quantum theory have been suggested.

The problem was brought into focus by a famous paper in 1935 by Einstein, Podolsky and Rosen, who argued that the strange behaviour of entanglement meant that quantum mechanics was an incomplete theory, and that there must be what came to be known as 'hidden variables' not yet discovered. This produced a famous debate between Einstein and Niels Bohr, who argued that quantum mechanics was complete, and that Einstein's problems arose because he tried to interpret the theory too literally.

However in 1964, John Bell pointed out that for certain experiments classical hidden variable theories made different predictions from quantum mechanics. In fact he published a theorem which quantified just how much more strongly quantum particles were correlated than would be classically expected, even if hidden variables were taken into account. This made it possible to test whether quantum mechanics could be accounted for by hidden variables. A number of experiments were performed, and the result is almost universally accepted to be fully in favour of quantum mechanics. Therefore there can be no 'easy' explanation of the entangled correlations. The only kind of hidden variables not ruled out by the Bell tests would be 'non-local', meaning they would be able to act instantaneously across a distance.

More recently, from the beginning of the nineties, the field of quantum information theory opened up and expanded rapidly. Quantum entanglement began to be seen not only as a puzzle, but also as a resource for communication. Imagine two parties, Alice and Bob who would like to send messages to one another over a distance. In 1993, Bennett et al. showed that if Alice and Bob each hold one of two particles which are entangled together, a quantum state can be transmitted from Alice to Bob completely by sending fewer classical bits than would be required without the entanglement. This process has been called 'quantum teleportation'. It involves not only bits for sending information, but 'e-bits', or entanglement bits, which consist of a maximally entangled pair of particles. Other ways in which entanglement can be used a an information resource have also been discovered, for example, dense coding, cryptography and applications to communication complexity. Entanglement was found to be a manipulable resource. Under certain conditions, states of low entanglement could be purified into more entangled states by acting locally, and states of higher entanglement could be 'diluted' to give larger numbers of less entangled states.

Investigation of quantum entanglement is currently a very active area. Research is being done on measures for quantifying entanglement precisely, on entanglement of many-particle systems, and on manipulations of entanglement and its relation to thermodynamics.

Quantum theory predicts that a vast number of atoms can be entangled and intertwined by a very strong quantum relationship, even in a macroscopic structure. Until now, however, experimental evidence has been mostly lacking, although recent advances have shown the entanglement of 2,900 atoms. Scientists at the University of Geneva (UNIGE), Switzerland, recently reengineered their data processing, demonstrating that 16 million atoms were entangled in a one-centimetre crystal. They have published their results in Nature Communications.

The laws of quantum physics allow immediately detecting when emitted signals are intercepted by a third party. This property is crucial for data protection, especially in the encryption industry, which can now guarantee that customers will be aware of any interception of their messages. These signals also need to be able to travel long distances using special relay devices known as quantum repeaters—crystals enriched with rare earth atoms and cooled to 270 degrees below zero (barely three degrees above absolute zero), whose atoms are entangled and unified by a very strong quantum relationship. When a photon penetrates this small crystal block, entanglement is created between the billions of atoms it traverses. This is explicitly predicted by the theory, and it is exactly what happens as the crystal re-emits a single photon without reading the information it has received.

It is relatively easy to entangle two particles: Splitting a photon, for example, generates two entangled photons that have identical properties and behaviours. Florian Fröwis, a researcher in the applied physics group in UNIGE's science faculty, says, "But it's impossible to directly observe the process of entanglement between several million atoms since the mass of data you need to collect and analyse is so huge."

As a result, Fröwis and his colleagues chose a more indirect route, pondering what measurements could be undertaken and which would be the most suitable ones. They examined the characteristics of light re-emitted by the crystal, as well as analysing its statistical properties and the probabilities following two major avenues—that the light is re-emitted in a single direction rather than radiating uniformly from the crystal, and that it is made up of a single photon. In this way, the researchers succeeded in showing the entanglement of 16 million atoms when previous observations had a ceiling of a few thousand. In a parallel work, scientists at University of Calgary, Canada, demonstrated entanglement between many large groups of atoms. "We haven't altered the laws of physics," says Mikael Afzelius, a member of Professor Nicolas Gisin's applied physics group. "What has changed is how we handle the flow of data."

Particle entanglement is a prerequisite for the quantum revolution that is on the horizon, which will also affect the volumes of data circulating on future networks, together with the power and operating mode of quantum computers. Everything, in fact, depends on the relationship between two particles at the quantum level—a relationship that is much stronger than the simple correlations proposed by the laws of traditional physics.

Although the concept of entanglement can be hard to grasp, it can be illustrated using a pair of socks. Imagine a physicist who always wears two socks of different colours. When you spot a red sock on his right ankle, you also immediately learn that the left sock is not red. There is a correlation, in other words, between the two socks. In quantum physics, an infinitely stronger and more mysterious correlation emerges—entanglement.

Now, imagine there are two physicists in their own laboratories, with a great distance separating the two. Each scientist has a a photon. If these two photons are in an entangled state, the physicists will see non-local quantum correlations, which conventional physics is unable to explain. They will find that the polarisation of the photons is always opposite (as with the socks in the above example), and that the photon has no intrinsic polarisation. The polarisation measured for each photon is, therefore, entirely random and fundamentally indeterminate before being measured. This is an unsystematic phenomenon that occurs simultaneously in two locations that are far apart—and this is exactly the mystery of quantum correlations.

Quantum Tunneling

The quantum tunnelling effect is, as the name suggests, a quantum phenomenon which occurs when particles move through a barrier that should be impossible to move through according to classical physics. The barrier can be a physically impassable medium, like an insulator or a vacuum, or it can be a region of high potential energy.

In classical mechanics, if a particle has insufficient energy to overcome a potential barrier, it simply won't. In the quantum world, however, particles can often behave like waves. On encountering a barrier, a quantum wave will not end abruptly - its amplitude will decrease exponentially. This drop in amplitude corresponds to a drop in probability of finding a particle as you look further into the barrier. If the barrier is thin enough, the amplitude may be non-zero on the other side, so there is a finite probability that some of the particles will tunnel through the barrier.

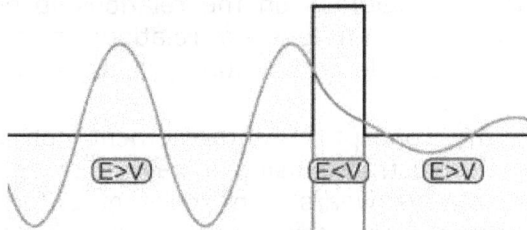

In regions where the potential energy is higher than the wave's energy, the amplitude of the wave decays exponentially. If the region is narrow enough, the wave can have a non-zero amplitude on the other side. Image Credit: Wikipedia

The tunneling current is defined as the ratio of the current density emerging from the barrier divided by the current density incident on the barrier. If this transmission coefficient across the barrier is a non-zero value, there is a finite likelihood of a particle tunneling through the barrier.

Discovery of Quantum Tunneling

The possibility of tunneling was first noticed by F. Hund in 1927, whilst calculating the ground state energy in a "double-well" potential - a system where two separate states of similar energies are separated by a potential barrier. Many molecules, such as ammonia, are examples of this type of system. "Inversion" transitions between two geometric states are forbidden by classical mechanics, but are made possible by quantum tunneling.

In the same year, L. Nordheim noticed another incidence of the tunneling phenomenon whilst studying the reflection of electrons from a variety of surfaces. In the following few years, tunneling was successfully used to calculate the ionization rate of hydrogen by Oppenheimer, and to explain the range of alpha decay rates of radioactive nuclei by George Garnow, and, independently, R. W. Gurney and E. U. Condon.

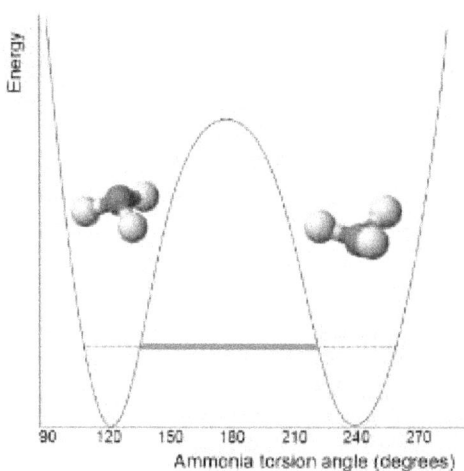

Ammonia torsion angle (degrees)

The two energetically equivalent states of ammonia, NH3, can exchange more readily than is classically predicted - this is because the molecule can tunnel through the potential barrier at lower energies than are required to pass through the transition state.

Quantum Tunneling in Nature

Although, like much of quantum physics, tunneling may appear to have little relevance to everyday life, it is a fundamental process of nature which is responsible for many things on which life itself is dependant.

It has even been hypothesized that the very beginning of the universe was caused by a tunneling event, allowing the universe to pass from a "state of no geometry" (no space or time) to a state in which space, time, matter, and life could exist.

Tunneling in Stellar Fusion

Fusion is the process by which small nuclei can join together to form larger nuclei, releasing huge amounts of energy. Fusion inside stars produces all the elements of the periodic table, except hydrogen, and fusion of hydrogen into helium is the process which gives stars their power.

However, fusion happens much more often than we originally thought it should. As all nuclei are positively charged, they repel each other very strongly, and their kinetic energy is very rarely sufficient to overcome this repulsion and allow fusion to occur. If tunneling effects are taken into account, however, the proportion of hydrogen nuclei which are able to undergo fusion increases dramatically.

This helps to explain how stars are able to remain stable for millions of years. The process is still not very likely though - an average hydrogen nucleus will undergo over 1000 head-on collisions before it fuses with another.

Tunneling in Smell Receptors

Until quite recently, it was believed that chemical receptors in the nose (400 different kinds in humans) detected the presence of various chemicals by a lock-and-key process, which identified the molecule's physical shape.

There are some issues with this theory, however. For example, ethanol and ethanethiol, which have very similar shapes, smell completely different (ethanol is the alcohol we drink - ethanethiol smells of rotten eggs). This suggests that some other identifying mechanism is at work.

A theory which has been growing in popularity over the last decade or so is that smell receptors rely in part on quantum tunneling to identify chemicals. The receptors pump a small current across the odourant molecule, causing it to vibrate in a characteristic way. In order for the current to flow, however, the electrons must tunnel through the non-conducting gap between the cells of the receptor and the molecule.

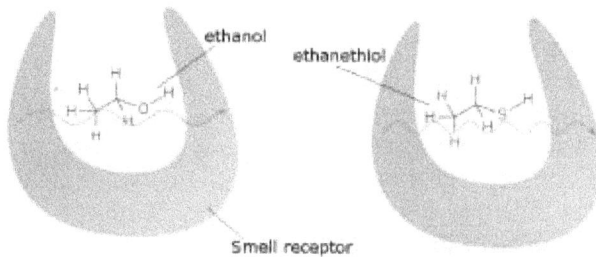

Smell receptors can detect the differences between similarly-shaped molecules by tunneling a small current across them, causing a characteristic vibration.

Applications of Quantum Tunneling
Josephson Junctions
Josephson junctions consist of two superconductors separated by a very thin layer of non-superconducting material, which can be an insulator, a non-superconducting metal, or a physical defect. The superconducting current can tunnel across the barrier, and the electrical properties of this system are very precisely defined.

This leads to several unique applications, primarily for highly precise measurements. Josephson junctions are common components in superconducting electronics, quantum computers, and are also used extensively in superconducting quantum interference devices (SQUIDs), which are capable of measuring extremely weak magnetic fields.

Tunnel Diodes
A tunnel diode is a semiconductor capable of high speed operation, comprising of a thin insulator sandwiched between

two semiconductors. Also known as the Esaki diode after L. Esaki for his work on the topic, this diode is capable of frequencies well into the microwave range.

Tunnel diodes are a two terminal device with a heavily doped p-n junction whereby the electric current through the diode is caused by quantum tunneling. Significantly, the current across the tunnel diode will decrease as the voltage increases meaning it will exhibit a negative resistance.

The concentration of impurities in a tunnel diode is 1000 times greater than a normal p-n junction diode, and as a result the p-n junction has an extremely narrow depletion region in the order of nanometres. In ordinary diodes, current is produced when an applied voltage is greater than the built in voltage of the depletion region.

However, in tunnel diodes, a small voltage which is less than the built-in voltage of the depletion region produces an electric current due to quantum tunnelling between the n and p region. The narrow depletion region is required so that the barrier thickness is low enough for tunnelling, as mentioned in the introductory section.

Tunnel diodes are used in logic memory storage devices, relaxation oscillator circuits and as ultra-high-speed switches. Owing to the high resistance to radiation, they are commonly used in the nuclear industry.

Scanning Tunneling Microscopes

A Scanning Tunneling Microscope (STM) works by scanning a very sharp conducting probe across the surface of a material. An electrical current is passed down the tip of the probe, and tunnels across the gap into the material.

As the gap gets wider or narrower, the tunneling current gets smaller or larger, respectively. Using this data, we can build an incredibly detailed picture of the surface, even to the point of resolving humps in the surface due to individual atoms. This technique has allowed leaps forward in our understanding of the physics and chemistry of surfaces.

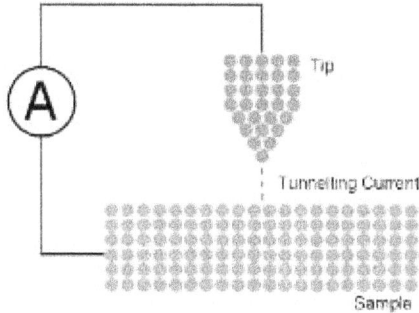

Schematic of an STM - the tunneling current varies with the distance between the tip and the atoms on the surface, allowing defects and even individual atoms to be mapped.

Flash Drives

Data on flash drives is stored in a network of memory cells made up of "floating-gate" transistors. These consist of two metal gates, a control gate and a floating gate. The floating gate is trapped in an insulating layer of metal oxide.

A floating-gate transistor in its normal state registers a "1" in binary code. When an electron is attached to the floating gate, it becomes trapped in the oxide layer, affecting a change in the voltage of the control gate - a transistor in this state registers a "0" in binary.

When data is erased from flash memory, a strong positive charge applied to the control gate causes the trapped electron to tunnel through the insulating layer, returning the memory cell to a "1" state.

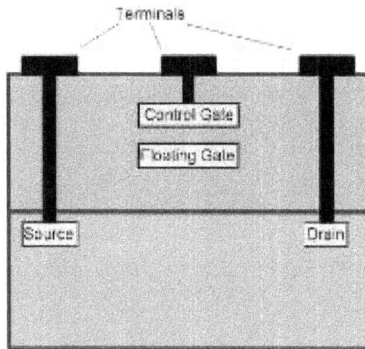

Schematic of a floating-gate transistor. Trapping an electron in the floating gate causes a change in the voltage across the control gate. Each transitor stores a bit of information.

The Speed of Quantum Tunneling

Tunneling is, at its heart, a wave phenomenon. Because quantum particles like electrons have wave nature, they can't come to a stop in an infinitely short time, but need to tail off a little more slowly, for basically the same reasons described in this animated explanation of the Uncertainty Principle: If you want a wave-like probability distribution to stop suddenly, you need to add a huge number of wavelengths to get that sharp edge, which means you would lose all information about its momentum. An electron with a reasonably well-known momentum, then, must have some uncertainty in position, which means the wavefunction will extend into regions the particle doesn't have enough energy to reach according to classical physics.

If you have a really large excluded area, the consequences of this are pretty negligible, since there's not much way to measure the presence of an electron inside a material that's designed to exclude it. Things get interesting, though, if you exclude the electron from only a narrow region of space, forming a thin barrier between two regions where it's perfectly happy to exist.

For a classical particle, the thickness of the barrier makes no difference-- if it can't enter, it can't enter and the probability of finding it in the forbidden region drops immediately to zero. The quantum particle, though, has a wavefunction that extends into the forbidden region. This drops off very fast as you move into the region where it shouldn't be-- it's generally an exponential decay-- but if the barrier is thin, the value of the wavefunction can be non-zero when you get to the other side. And that means there's some probability to find the particle on the far side of the barrier, even though it should be impossible for it to get there.

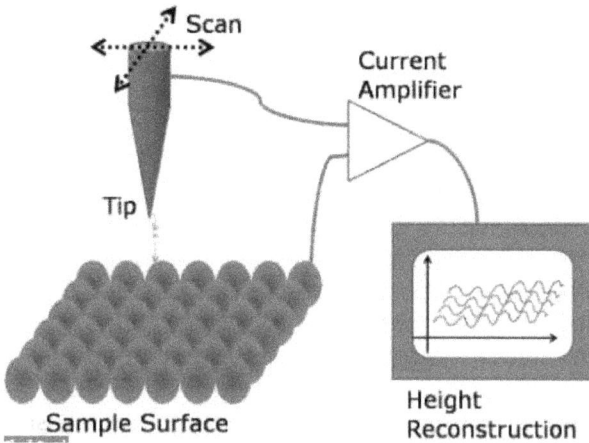

Schematic of a scanning tunneling microscope: a sharp tip is scanned over a surface, and the current from electrons tunneling to the surface lets you reconstruct the height.

This is an odd phenomenon, but it has enormous practical applications. For one thing, it makes it possible for the Sun to shine. It's also the basis of the Scanning Tunneling Microscope, which uses the exponential decay in the probability of electrons tunneling across a tiny air gap as an incredibly sensitive probe of the distance between a probe and the surface of some sample. This is the technology that allows scientists to make images of matter that resolve individual atoms and molecules sitting on some surface.

So, tunneling is unquestionably a real process, which then raises the question of exactly how this business works. That is, what happens to the particle as it's crossing the barrier, and how does the crossing affect its properties? Specifically, is there a measurable effect of time spent in the crossing process?

These are really difficult questions to answer, because the time scales are so short. Whenever you start talking about time and distance in physics, you can define a sort of natural time scale, namely the time needed for light to cross the relevant distance. With tunneling, you're dealing with really small distances. The time required for light to cross the thickness of an atom (which is the relevant scale for an STM) is less than an attosecond (0.000000000000000001s, if you like to see lots of decimal places). If you're going to look for a time delay associated with tunneling, you need a way of timing your experiment at that kind of level, which is really hard.

In recent years, people have been investigating this process using a "streaking" technique with ultrafast lasers. They don't directly measure the start and end of the tunneling process, but they use a clever analysis trick to set up a situation where they can measure a change in trajectory for electrons tunneling through a barrier.

The streaking technique starts with a sample of atoms, which they blast with a super-intense laser pulse that lasts only a few femtoseconds. This pulse is sufficiently intense that it doesn't make sense to talk about in terms of unimaginably huge numbers of photons; instead, it acts like a huge electric field, which lowers the energy for an electron on one side of the atom and raises it on the other. The shift is almost but not quite enough to rip an electron out, but leaves a small barrier holding it in.

Of course, any time you have a small barrier and a quantum particle, you get tunneling, so even though the pulse technically isn't strong enough to ionize the atoms, they nevertheless see some free electrons produced. These escape via tunneling, and because tunneling decreases exponentially with the size of the barrier, the vast majority of the tunneling should occur in a really short window around the time when the pulse is at its most intense. This gives a reasonably well-defined "start" time for the process.

Once the electrons are out, they get pushed around by the same giant electric field that extracted them, and this is where the "streaking" part comes in. The exact direction of the push on the escaping electron depends on the orientation of the field. The clever trick they use to get timing information is to use a circularly polarized laser pulse, where the direction of the electric field rotates at the speed of oscillation of the laser. The direction of the initial push, then, depends on exactly when the electron comes out-- if it emerges instantaneously, it starts off pushed in one direction, but if the tunneling requires some time, then the light field has rotated, and it gets pushed in a slightly different direction. They use a position-sensitive detector to measure exactly where the electrons end up, which will depend on that initial push.

This is, of course, a ferociously complicated process, with lots of fiddly details that could confound the measurement. Which is where the argument comes in: different groups use different methods to try to sort this out, and come to different conclusions about the tunneling time.

In the Physical Review Letters paper linked above, they opt for a differential technique: rather than trying to take on the nearly impossible task of locking all those factors down, they look at a difference between two very similar atoms: argon and krypton, which are mixed together in their target. Argon and krypton sit next to each other in the noble gas column of the periodic table, which means that they're extremely similar chemically. The only difference between them that really comes into the current experiment is the ionization potential: it takes slightly more energy to pull an electron off argon than krypton.

This difference in ionization potential shows up as a slight difference in the height and thickness of the barrier the electrons have to tunnel through. Electrons leaving argon need to go a little farther than electrons leaving krypton, which means that if tunneling takes time, it should take them a little longer to make their way out. This manifests as a small difference between the position distribution for electrons from argon atoms and electrons from krypton atoms, and that's the thing they measure.

This is a clever trick, because everything else is the same-- the same detector, the same laser pulse, etc. So they take out most of the possible confounding factors, and get a difference that they can compare to theoretical predictions. When they do that, they find that their results are consistent with the picture where electrons need some time to tunnel through the barrier, and not easy to explain in the picture where they just instantaneously appear on the far side.

So, what's the problem? Well, the "compare to theoretical predictions" step is a bear, because argon and krypton are fairly complicated atoms. So, another experiment, posted to the arxiv a few days after the paper linked above was published tries a different attack: They work with hydrogen, which is the only atom simple enough to allow exact calculations.

The core technique is the same: ultrafast intense laser pulse, circular polarization, streaking detection. The difference is that the simplicity of hydrogen makes it easier to calculate the expected trajectories for the comparison. When they do that, they claim that there's nothing in the data to indicate a finite tunneling time. They see a variation in the angle depending on the laser intensity (which varies the height of the barrier through which the electrons must tunnel), but they claim this is due not to the timing of the tunneling, but the electrostatic tug of the proton left behind. They try to verify this by re-doing the calculation in a way that shuts off the force from the proton, and find no change in the trajectory with laser intensity. This, they say, indicates that the "tunneling time" can't be any longer than a couple of attoseconds.

On the one hand, the differential measurement comparing argon and krypton should remove most of the complications that make the theory hard to do for multi-electron atoms. On the other, though, they're still very much not hydrogen, and I believe theory calculations involve hydrogen a lot more than those involving argon and krypton... There will inevitably need to be further experiments refining and expanding the technique before anything is completely settled, and if you read the papers closely, there's a bit of sniping in there indicating that this is likely to be a lengthy and contentious process.

*****

Epilogue

In statistical mechanics, Boltzmann's formula means that the entropy S of an isolated system at thermodynamic equilibrium is defined as the natural logarithm of W, the number of distinct microscopic states available to the system given the macroscopic constraints (such as a fixed total energy E):

$$S = k \log W$$

This equation, which relates the microscopic details, or microstates, of the system (via W) to its macroscopic state (via the entropy S), is the central idea of statistical mechanics. This entropy exactly corresponds to Shannon's information entropy.

Let us see how:
Suppose there are 6 colored balls and you are asked to select two balls out of them. The number of ways that you can select two balls out of 6 is $^6C_2$ Which equals to 15. So W=15.

$$\text{Boltzmann's entropy } S = k \log 15$$

Shannon's entropy formula is : $S = -\Sigma p_x \log p_x$ where p_x is equal to the probability of any permutation out of the total number of permutations.

Since there are 15 ways of selecting 2 balls out of 6, the probability for any permutation is 1/15 and there are 15 such permutations. If we inject these values into Shannon's entropy formula –
$S = -\Sigma p_x \log p_x = -[(1/15) \log (1/15) + (1/15) \log (1/15) + (1/15) \log (1/15).....]$ upto 15 times
$= -15[(1/15) \log(1/15)]$
$-\log(1/15)$ which is equal to $\log 15$.

So , Shannon's entropy $S = \log 15$ which is matching with Boltzmann's entropy.

From the above formulas , it is clear that entropy is somehow related to information.

The entropy of a system equals to the randomness of its constituent parts. The more random it is, the more entropy there is.

The more the entropy of a system, the more information it contains. Information always passes from sender to receiver. The more entropy the sender has, the more unpredictable it is and the more information it has. But from receiver's point of view , once it receives the message, it should decrease the unpredictability, so the less entropy the receiver has.

So from receiver's perspective, the more entropy , the more uncertainty it has about the system and the less information it has consequently.

*****

www.ingramcontent.com/pod-product-compliance
Lightning Source LLC
Chambersburg PA
CBHW060402200326
41518CB00009B/1229